He Said, She Said

The Tesla Model 3 User's Guide

Get Mansplained and Ma'am-splained all in one book

Sheryl Scarborough & Jerry Piatt

PS PROD LLC

ISBN: 978-1-7342153-1-1

Cover design by Brett Lyerla
https://www.brettlyerla.com/

Disclaimer:

Neither the authors nor the publisher are in any way experts on the topics discussed in this e-book. The authors only wish to share their own experiences learning about driving their Tesla with you. The information in this book is designed to provide helpful information on the subjects discussed. It is intended for informational and entertainment purposes only and not for expert advice. **Professionals should always be consulted as needed for questions or problems about any of the actions described or recommended within.**

Limit of Liability/Disclaimer of Warranty: The authors and the publisher make no representations or warranties with respect to the accuracy or completeness of the contents of this work and specifically disclaim all warranties, including, without limitation, warranties of fitness for a particular purpose. The advice and strategies contained herein may not be suitable for every situation. Estimates, graphs, and tables are speculative and should not be considered appropriate or suitable for planning purposes. The fact that a product, organization or website is referred to in this book does not mean that the author or publisher endorses a product, or the information or the recommendations of the organization or website. Readers should be aware that the websites listed in this book may change. Neither the authors nor the publisher shall be held liable or responsible to any person or entity with respect to any loss or incidental or consequential damages caused, or alleged to have been caused, directly or indirectly, by the information contained herein.

Errors and omissions: The information presented herein is accurate to the best of the authors' knowledge at the time of writing. All Tesla vehicles and the Tesla supercharger infrastructure are evolving rapidly. It is possible that factual information about Tesla products and the infrastructure that supports them may have changed since the publication of this book. Always refer to the Tesla website, official Tesla printed documents, and Tesla representatives for the most up-to-date information on a vehicle, its options, and pricing.

Stock Ownership Disclosure: As of the date of launch of this e-book, Sheryl and Jerry own 75 shares of Tesla stock. This is a disclosure and not a recommendation. Consult a professional financial advisor for any investment decisions.

Dedication:

For Elon...Stuart...Greta
and the battle for a carbon neutral future

Table of Contents

INTRODUCTION

"The first step is to establish that something is possible; then probability will occur."
—Elon Musk

SHE SAID:

If you're a new Model 3 owner, CONGRATULATIONS! If you're only Tesla-curious, welcome. This book should offer something for both of you.

Owning a Tesla Model 3 puts you in a whole new club of happy people, zero emissions and car-love like you've never experienced before. There's new terminology and a host of cool, new drive time features. But mostly what I've come to understand is that regular car drivers have no idea what goes on inside my Tesla.

And by inside, I don't mean under the hood.

My car has the ability to drive me! She remembers my preference for *chill mode with confirmation* which is different from my husband's strategic *Mad Max* settings. My car learns the routes I routinely travel and improves the self-driving of them over time. My car is connected to all the other Teslas on the road and through that data, she gets better and smarter every day. The many ways to customize a Model 3 are why my husband, Jerry Piatt and I are the perfect choice to write a user's guide and put them all in one place. Follow along with me here...

- I write mysteries...is there anything more mysterious sometimes than technology?
- Jerry Piatt is a retired software engineer who doesn't believe technology is mysterious at all. Maddening at times, sure. But mysterious? Nope. Not really.
- Being a guy, Jerry knows a lot about cars.

- I have driven a lot of cars and still swear that magically banging my high heel on the carburetor of my 240Z got it running again. You believe me, right?
- I know how to outline, research and pull a book together. I've been paid to do it for other companies. With Jerry as my resident expert, anything that clocks in above my explanatory pay grade will land squarely in his. Plus, Jerry recently retired so he needs something to do...right?

HE SAID: *Funny, considering the Honey-Do home improvement list she's given me.*

I can't tell you how excited Jerry was about this project. I can't tell you because he wasn't excited. Not in the least.

"Write a book?" he said with a scowl. "You mean like the manual?"

Yeah. Okay. He had a point.

And yet it's clear from the multitude of forum posts I've read we may all have access to the Tesla Owner's Manual but there's really no substitute for that information coupled with real-world experience.

This guide (which is half the size of the manual) is comprised of well-researched data and anecdotes (our own, plus stories and concerns gleaned from miles of forums). We even include some of the more infamous Tesla Model 3 legend and lore. There are things that will be obsolete before this book even gets through formatting. That's a basic truth of technology and we accept that. We're intrepid. We love our Tesla and believe in the company and Elon's mission. Also, our non-Tesla-owning friends were tired of hearing us enthuse over it and we needed somewhere to focus all that energy.

The Tesla Model 3 is primarily software driven. This means nearly every part of your driving experience can be configured to your personal preference. And, while there's a ton of great Model 3 information on YouTube and various forums, finding out about a feature you didn't even know you wanted or needed is sometimes easier in a book.

So, a mystery writer and software engineer buy a Tesla and voila! You get a book.

We opted for a **HE SAID/SHE SAID** approach because, while we love the car the same, our individual experiences *are* different. He speaks deep-geek while I offer, *don't-worry-about- that...just-do-this* advice. We spent weeks pouring over the Model 3 Owner's Manual, combing various Tesla forums and owners' groups, tracking down articles from auto journals and business journal op-eds, not to mention viewing countless YouTube videos. We even ran specific tests on our own car. We did all of this so you don't have to.

You're welcome!

Now what are you waiting for?
C'mon. Plug in! The future is here and it's amazing!

HOW TO USE THIS GUIDE:

Some of the information we want you to have can only be passed along in the form of a link. In the book we have expressed these CHECK OUT links in such a way that typing them into a search bar will take you to the information or video. Additionally, all the links mentioned in the book can be found on our website: www.TeslaModel3Guide.com/book-links. This is also where we will address software updates as they happen. We'd love to hear from you, so please email your questions, comments, requests and photos of your awesome vehicles to: HeSaid@TeslaModel3Guide.com or SheSaid@TeslaModel3Guide.com.

SHE SAID:

Her name is Helva-Pearl Piatt and she became mine on January 19, 2019. (Note: if I could put little hearts and stars around the word MINE I would!) She's a white-on-white, LR AWD Tesla Model 3. I opted for Enhanced AP at time of purchase and bought FSD when it went on sale. At the time all the Model 3s came with premium everything. In car parlance, she's fully loaded!

THIS MEANS:

Helva-Pearl is a dual motor, all-wheel drive Model 3, with a long range 75 kWh battery, a driving range of 310 miles and all the bells & whistles. She can go from 0-60 in 4.4 seconds! She has a pearl white exterior with a white "vegan" leather interior. And she is gorgeous! Make no mistake. I love Pearl like a nine-year old loves cake!

10 THINGS TO LOVE ABOUT THE TESLA MODEL 3

1. She's beautiful. I may not turn heads anymore, but she does! I think of her as my exoskeleton.
2. She's fast. Low profile. Bottom heavy. Corners like a dream. #sportscar
3. She's smart and features a level of automation that will make your head explode.
4. Every single part of her is engineered to be the very best in her class. Down to and including the Aero Wheel covers, which on the attractiveness scale are more an automotive Birkenstock than a Louboutin. But hey, engineers FTW!

   HE SAID: I wanted the 19" Sport Wheels but she said why are we doing this again...to save the planet or to drive fast? It's her car. So, what could I say?

5. She's safe...but also FAST!
6. She's extremely easy to maintain. For the sake of my husband's sanity something needed to be low maintenance in this household. #nailedit #justsaying #TeslaFTW
7. The AP features are the best thing about her. The self-driving really reduces driving stress. Don't buy into the negative press, NO ONE does this better, or even close to Elon!
8. Great sound system. Super easy to pair and program with my phone and the scroll buttons on the steering wheel are surprisingly handy.
9. Two words: TESLA RANGERS, the mobile service and repair fleet that comes to you!
10. NO GAS STATIONS! I used to love gas stations so much I'd bring our Prius home on fumes and mention needing gas to my husband. (See item 6). Extra points for night before early morning meeting! So yeah, zero emissions. Now, I plug in at home, in my garage. My car is juiced up and ready to go every morning. She's lean, clean & mean...also, did I mention she's FAST?

HE SAID:

1. Retired people are busy. I don't have time to run around getting oil changes and radiator flushes, etc. Other than tire rotations, the Model 3 requires scheduled maintenance only every two years. At some point, it may be possible to just have the car drive itself to the service center and return when the service has been completed. Elon? Are you listening?

2. I'm tall. My wife, not so much. Seat and mirror positions are no longer a point of contention. In fact, there are many driving options that are automatically set to my preference just because I got in the car on the driver's side. The only thing I have to do is adjust the rear-view mirror. Every other customization is set automatically. Boo-Ya!

3. Wintertime: No more freezing my nards off while waiting for my car to warm up so I can turn on the heater. I can pre-heat my Model 3 (including the seats) from ANYWHERE using the MOBILE APP. So nice to get into a toasty warm car on a cold winter morning.

4. Summertime: Ever been running errands and had to alter your route so you pick up groceries (or plants, animals, your mother-in-law) last so they don't bake in a hot car while you're making other stops? You can instruct the Model 3 to leave the air conditioning on when you leave the car, then monitor and even adjust the temp from the Mobile app.

5. Sports cars sprint out and corner really nicely, but the suspension lets you feel every bump in the road. Luxury cars have a smooth ride but seem to take forever getting up to cruising speed. Imagine a four-thousand-pound sports car with all the weight at the very bottom. Now picture this car, lunging from 0-60 MPH in 3.2 seconds and running up to a top speed of 162 MPH. That's the Performance version of the Tesla Model 3. Low center of gravity means it corners better than a mid-engine V8 sports car. Full

disclosure—our LR AWD does 0-60 MPH in 4.4 seconds and the top end is only 145 MPH. I make no apologies.

SHE SAID: I think expressing speed in 0-60 MPH/seconds is more meaningful to guys. So, ladies, here's my description of the Model 3's impressive power and acceleration. When I demonstrated it for a friend on a freeway on ramp, I punched it and the force blew my sunglasses off the top of my head, clear into the backseat! #power4days

6. Sentry Mode security FTW! Don't mess with my Tesla. I will record you and turn it over to the Po-Po. Sentry mode alerts you via the MOBILE APP if someone tampers with your car AND it actually records the activity all around the car, capturing the last ten minutes of video prior to the attack.
7. I love the **Navigate-On-Autopilot** feature. Look, ma! No hands! (almost). You DO have to keep your hands on the wheel (kinda) and you need to pay attention, but you'd be surprised how relaxing it can be to have the car remember the route and make all the necessary lane and route changes for you.
8. **Traffic-Aware Cruise Control** is part of **Autopilot** but, used by itself doesn't autosteer or change lanes automatically. Still, its ability to manage the ebb and flow of traffic speeds without my having to move my foot from accelerator to brake and back constantly makes freeway driving downright relaxing—regardless of traffic.
9. No more brake dust on my wheels. That fine material that comes off the brake pads when they make contact with the rotors to slow the car? You'll likely never see this on a Model 3. Nearly all braking is regenerative braking. Regenerative braking doesn't involve the disc brakes at all. It uses the motor/generator(s) to slow the car and, in so doing restore some charge to the battery.
10. A LOT of the features in the Model 3 are implemented in software. Sometimes there are bugs. Sometimes the engineers get ideas for new features. Both the bug fixes and any new features are delivered directly to the car while it's parked and connected to Wi-

Fi. These OTA updates are part of the reason you will rarely ever need to take your Model 3 to an actual Tesla Service Center.

TERMINOLOGY:

AP	Autopilot self-driving feature. Currently included on all Model 3s
AVATAR	(in this context) An animated, on screen representation of your Model 3
AWD	All-wheel drive
CARD READER	there are 3 RFID card readers on the car which read the Tesla key card
CURB RASH	An unsightly scrape on the rims caused by parking too close to a curb.
DOG MODE	A climate control setting to keep the temperature inside the vehicle, while parked, to a level of comfort for a dog, but humans use it too!
DUAL MOTOR	Two independent motors digitally control torque to front and rear wheels. Dual motors respond to changing conditions in as little as 10 milliseconds.[1] (Otherwise known as 1/100th of a second). Also, if one motor stops working, the car can safely continue to drive on the other.
ENERGY APP	An APP located on the Model 3 touchscreen that predicts range requirements in advance of a trip, then tracks that prediction on a graph against the real time, actual usage. Invaluable for road trip monitoring.
EV	Electric Vehicle, means fully electric not hybrid or EV combined with gas.
FRUNK	The front trunk or space where the engine resides in an ICE vehicle. In the Model 3 this = extra cargo space.
FSD	Full self-driving (a future feature not yet employed but one which Elon has promised will allow an owner/driver to input an address into

	the navigation instructing the car to drive on its own, door-to-door, from one location to another without human intervention.)
FTW	Acronym: For the win!
GIGAFACTORY	A lithium-ion battery and EV sub-assembly plant owned and operated by Tesla. Plant 1 is located near Reno, Nevada.
HYBRID	A dual fuel automobile. There are various combinations and configurations which combine gas + battery. (NOTE: if the car requires gas it will also require all of the other mechanical parts of a traditional ICE vehicle which defeats the low maintenance benefits of an EV.)
ICE VEHICLE	ICE = Internal Combustion Engine. (Remember those?)
IDLE FEES	Charges which might apply to your Tesla account if you remain plugged into a Supercharger after your car is fully charged.
JOE MODE	A setting that reduces the volume of the Tesla chimes to avoid waking up sleeping children in the backseat. (Access from DRIVING PREFERENCES menu > SAFETY & SECURITY > JOE MODE.)
J1772 adaptor	The standard North American electrical connector for charging EVs. (Tesla uses a proprietary connector for their Supercharger network but a J1772 connector came with our car expanding our charging options.)
JUICE OUT	(slang) Running the battery down to zero. Not a good thing to do.
kWh	Kilowatt hour as a measure of battery capacity or state of charge.

LR	Long Range Abbreviation for Model 3 configuration that includes the largest battery capacity (75kWh).
MOBILE APP	Tesla App for iPhone or Android that will control your Model 3.
MOTOR	Electric vehicles have MOTORS not engines.
NEMA 5-15	Standard 110 VAC outlet (i.e. regular outlet)
NEMA 14-50	220 VAC outlet (i.e. electric clothes dryer outlet)
NOA	Navigate on Autopilot (see **Chapter 5: Bells & Whistles**)
OTA	Over The Air updates. (So. Much. Fun!)
RANGE	Distance the car can travel on a fully charged battery.
RANGE ANXIETY	The (usually) unnecessary fear of running out of battery before reaching a charging location.
REFERRAL CODE	A Tesla promotion to earn FREE Supercharger miles. (Provide a referral code from a Tesla owner when you order your Tesla and BOTH of you earn free Supercharging miles.) **Our REFERRAL CODE: Sheryl Scarborough or go to https://ts.la/sheryl51287**
REGENERATIVE	pertaining to Brakes. The motors can put charge back onto battery when you slow down. It's magic!
ROLL COAL	A huge plume of black or gray exhaust fumes, emitted by a modified diesel engine, which engulf the vehicle (usually a Tesla or Prius) traveling behind it. It seems this is how some diesel truck drivers get their jollies. (**SHE SAID:** *I've seen the videos. It's nasty. Go ahead and Google it.*)
RTFM	Acronym for Read The Farking Manual

RWD	Rear wheel drive (single motor, in the rear).
SENTRY MODE	A built-in security feature that uses the exterior cameras on the body of the car to record someone trying to break into or vandalize your vehicle.
SoC	State of charge or charge level of the battery.
SR	Standard Range Abbreviation for Model 3 configuration that includes the standard battery capacity (50kWh).
SUMMON	The ability to stand outside your vehicle and *cause* it to drive into or out of a parking space or garage with only the push of a button on the Mobile App or key fob.
SUPER-CHARGING	Always refers to the Tesla Supercharger network.
SWIPE	With fingertip SWIPE UP, DOWN or RIGHT or LEFT to OPEN or CLOSE menus on the touchscreen.
TAP	Quick tap on an ICON with fingertip to interact with touchscreen.
TACC	Traffic-Aware Cruise Control (aka: a commuter's answered prayers!)
TESLA RANGERS	Mobile fleet of mechanics that will come to you in specially outfitted trucks or Model S cars. They can handle 75% of all repairs at your location.
TIPPING POINT	"The critical point in a situation, process, or system beyond which a significant and often unstoppable effect or change takes place."[2]
TOW MODE	A setting for pulling a Tesla onto a flatbed tow truck.

Wh/mi	Watt hours per mile. (see **Chapter 8: Geek Stuff** for the full, head-exploding explanation).
V10	Software version 10, announced in July 2019, promised in August 2019 but dropped last week of September 2019 (as we were going to press).
VALET MODE	A built-in driver profile which limits speed, disables voice commands and Autopilot convenience features.[3]
VAMPIRE DRAIN	Range loss when car is parked and not plugged in. (**Chapter 3: Charging**)
VOICE CONTROLS	Built-in microphone for phone calls, navigation and playing music.

1

EV vs ICE

"I always have optimism, but I'm realistic. It was not with the expectation of great success that I started Tesla or SpaceX... It's just that *I thought they were important enough to do anyway."*

—Elon Musk

HE SAID:

When I was a young guy, I took pride in being able to do some minor maintenance on the cars I owned. Okay, I may have caused more problems than I fixed but I did learn to change my oil, points, plugs and condenser. This was a LONG time ago. Eventually, it became a chore that I just didn't have time for so I started paying to have it done. Then it became an expense. But it was always something I had to monitor and set up. Is it time for an oil change yet? Am I due for other maintenance (transmission fluid change, radiator flush, etc.)? I always worried I'd forget a key maintenance milestone. Oh, it happens! My oldest son blew the engine on his first car because he failed to change the oil when it needed it. Our youngest dropped his transmission because same thing with the transmission fluid. Expensive mistakes but they happen because it is difficult for busy people to keep track of and make time for all the maintenance requirements of an ICE vehicle.

Internal combustion engines (ICE) are intricate and complicated mechanisms. According to a June 2016 article on **CNBC.com** electric vehicles will soon be cheaper than regular cars partly because maintenance costs are lower.[4]

ICE vehicles typically have more than 2,000 moving parts while EVs have about 20. That's 99% fewer parts! This is the most compelling evidence for EVs to replace gasoline powered cars in the very near future. You don't have to be a tree hugger to appreciate the reduction in cost and hassle of driving a vehicle that requires virtually no maintenance.

Imagine never having to take your car in for an oil change and lube. Not having to worry about periodically flushing the radiator or changing the transmission fluid. You will want to have your brakes checked at reasonable intervals, of course, but some Tesla drivers report driving over 100,000 miles without having to replace brake pads. (Nearly all braking on a Tesla is regenerative braking which doesn't involve the wheel brakes at all.) No fuel injectors to clean, spark plugs to replace or timing chain to break. What will you do with all the free time and empty space on your calendar, not to mention the saved $$$?

CHECK OUT: **eGallon** Comparison cost of EV charging by state:

www.energy.gov/maps/egallon

Here's the list of required maintenance for your TESLA Model 3:

- Rotate the tires about every 10,000 – 12,000 miles.
- If you live in a snowy area where they use salt on the roads, the brake calipers should be cleaned and lubricated yearly or every 12,500 miles.
- Check the brake fluid and replace the cabin air filter every 2 years or 25,000 miles.
- Replace the battery coolant every 4 years or 50,000 miles.
- Replace the A/C desiccant bag every 6 years.

That's it, unless you count keeping an eye on your tire pressure and topping off the windshield wiper fluid when needed. You'll receive a message on the touchscreen when wiper fluid is low.

SHE SAID:

He's the tech-pro in our family, but I'm the one who drives our tech acquisitions. I am that person who will ~~waste~~ spend a half-a-day figuring out how to format and print labels on my computer when I could hand-address them in 15 minutes. If I can off-load *any* of my duties for driving or maintaining a vehicle to technology—no matter how small—I'm IN.

I upgraded to a Tesla Model 3 from a Prius, which was a technological and engineering marvel of its time. It was designed, from the ground up, to save gas. And it did. The Prius was so successful it inspired the hybrid movement. And yet none of the other hybrids ever exceeded the Prius in MPG without giving up something in the way of size, weight or comfort. In my opinion, this is because the other auto manufacturers didn't care about being the best, they just wanted to say they had one. By being first, Toyota had to care. They had to be the best or the hybrid experiment would have failed.

I see the same mindset with Tesla. The Tesla Model 3 was designed, from the ground up to be a safe, low maintenance, long-running, efficient, mass produced, electric car. I remember when Elon Musk first announced he was going to make electric vehicles. Laughter rippled through the auto industry. All the car articles (car-ticles?) said: "He'll never get a battery with enough range...and, if he does, it won't be a car anyone wants to drive."

Elon said: "Hold my beer!"

As of this writing, Tesla EVs have longer range than any other EV currently on the market. The Model 3 was named "most satisfying car" on the market by Consumer Reports, based on the results of surveys of more than half a million car owners.[5] And, despite what Wall Street shorts want you to believe, Model 3 demand AND sales are through the roof.

BUT THEY'RE SO...NEW:

It's a little scary to be an early adopter because the next version of any tech product will always be better, faster, cheaper—and probably all of the above. But if you're considering purchasing an EV or even just EV curious, here are the facts.

TESLA is in its TENTH year of manufacturing and selling electric vehicles.

There are SIXTEEN different EV models currently on the market and more are coming soon.

We're not alone in the transition to EV autos. More than one million plug-in passenger cars and vans have been registered in Europe. 1.6 million EVs have been sold in China. It is not a passing fad. The tipping point for EVs has been achieved. For the first time ever, I foresee a future where the average car I pass on the road will be an EV.

The question is when will *you* get onboard?

Not if...but when.

2

RANGE

"When somebody has a breakthrough innovation, it is rarely one little thing. Very rarely, is it one little thing. It's usually a whole bunch of things that collectively amount to a huge innovation."

—Elon Musk

1. How far can you really go?
2. How long does it take to charge?
3. How much does it cost?

These are the top three questions asked about our Tesla. Here are the quick answers.

Our LR, AWD Model 3 has a top range of 310 miles, but this distance is mitigated by weather, road conditions, elevation and driving practices (speed), coupled with the use of AC or heat. Real world: a 250-mile drive is easy on a single charge, without having to make a charging stop.

- 15 to 40 minutes at a Tesla Supercharger. 3+ hours on a home charger or overnight. (See Chapter 3: Charging for more details.)
- Supercharger fees (at time of this writing) are about $ .26 per kWh or $19 to fill our 75-kWh battery to 100%. To calculate home charging cost, you need to know the cost per kWh in your area and the size of your battery. (Ours: $ .07 per kWh and a 75-kWh battery.) Or $5.25 to fill battery to 100% at home. (NOTE: you rarely fill the battery to 100%. And you are never really charging from zero percent.)

HOW MUCH RANGE DO YOU NEED?

HE SAID:

Since I retired, I don't drive as much and my wife works from home as a writer so, between us we don't need a lot of range. But we're atypical. Let's talk about you.

The average American commute is 16 miles[6]. The average American drives a total of about 37 miles a day.[7] Most Americans can easily get through the day without worrying about charging, even if they own the Standard Range Model 3, which has an Advertised EPA Rated Range of 240 miles. In fact, tests show if you drive this car at 55 MPH you could get up to 304 miles on a fully charged battery. Of course, if you drive like Jimmie Johnson or Courtney Force that number's going to change in the other direction.

You monitor your State of Charge using the battery icon (located below and to the right of the speed indicator on the Driving Pane of the touchscreen). This either displays the percentage of charge or the estimated number of miles left in the battery. You can choose which setting you prefer from the DRIVING PREFERENCES menu > DISPLAY > ENERGY DISPLAY > ENERGY.

The Tesla Range Table at teslike.com/range.

It features continuously updated data on range and driving conditions for Tesla batteries.

WHAT ABOUT ROAD TRIPS?

When I mention traveling in an EV to my ICE-driving friends, the first thing they ask is, "What if you're out in the middle of nowhere and you run out of charge?" This earns a sigh, an eyeroll, maybe some head shaking. "What happens if you're out in the middle of nowhere and you run out of gas?"

Model 3 owners refer to this paranoia as *range anxiety*. I should say *new* Model 3 owners do because, after taking a couple road trips, all Model 3 owners quickly realize there's no reason for concern if they plan their trip and monitor their progress. I mean, you wouldn't take off into the Mohave

Desert in an ICE car without figuring out how many miles you're going to drive and whether you'll have enough fuel to get to the next gas station, right? It's no different with a Model 3.

WHY IT'S GREAT...

Tesla has a really good network of Superchargers (over 14,000 worldwide and expanding) that makes long distance travel easy and stress-free. In addition, there are EV chargers operated by independent companies like EVgo, ChargeHub, ChargePoint and Volta all over the place these days. In a pinch, nearly all of them can be used to charge your Model 3 using the J1772 adaptor that comes with your car. Tesla Superchargers will charge your car MUCH faster and probably cheaper than the generic chargers though, so plan on using them whenever possible.

PRO TIP:

Download the apps for the third-party EV networks to your smartphone and set up your credit cards in advance so they'll be ready to go if you need them. Any charges incurred on the Supercharger Network will be applied to your Tesla account and charged to the credit card you have on file there.

ROAD TRIPS: IMPORTANT THINGS TO KNOW

Speed affects range. The difference between driving at a sustained speed of 80 MPH versus driving at 65 MPH is about 50 miles of overall charge range[8].

In my experience, rain affects mileage, as well. I haven't seen an official explanation for this, but some feel it has to do with reduced road traction.

If you are using NOA, you will get a warning on the touchscreen if your speed presents a problem in reaching your destination. It will say something like: "To reach your destination, drive below xx MPH." DO NOT ignore this message. Either slow down or find a closer destination where you can add some charge.

When a Tesla Supercharger location appears on the touchscreen map, you can TAP it to display a small legend which tells the address, how many chargers are currently available, the maximum rate of charge (kW) available, and the price per kW for charging and for idle fees. It will also offer a NAVIGATE button, allowing you to immediately set a course in your GPS navigation to this location.

Tesla's online trip planner will help you plan your route and show you where you should stop to charge along the way. Good for night-before planning.

GPS navigation on the touchscreen will do the same in real-time and is what you will probably use when actually on the road.

The ENERGY APP (accessed from the APP LAUNCHER toolbar) provides a graph that updates in real time tracking your actual energy consumption against the projected use based on the navigation parameters for your trip. The Energy App predicts what percentage of charge will remain when you arrive at your proposed destination. This is the most accurate way to monitor your progress against the need for an extra charging stop or adjustment to your speed.

A free, third-party app called A BETTER ROUTE PLANNER provides even more detailed driving and charging assistance and can be accessed from the web browser on the touchscreen (www.abetterrouteplanner.com).

3

CHARGING

"You shouldn't do things differently just because they're different. They need to be...better."
—Elon Musk

THE COMMON WISDOM AT THIS TIME

- Only charge to 100% right before a road trip.
- Charging to 90% is okay, per Elon.
- Keep your Tesla plugged in at home, at night.

CONNECTORS

What to keep in the car and what to leave home?

Our Model 3 came with a charging cable and two adapters. One is useful for charging from a standard 110 VAC outlet (NEMA 5-15). The other can be used to charge from a 220 VAC outlet (NEMA 14-50). This would be the kind of outlet used to power an electric clothes dryer or a 220V welder. In addition, we received a J1772 adapter which can be used to connect almost any EV charger to your Model 3.

In normal circumstances, you will only need to travel with the J1772 adapter. However, for long trips, it may be advisable to pack the cable and both the 120VAC and 220VAC adapters. In the unlikely possibility that you require a charge and don't have enough SoC left to make the next Supercharger, you might be able to convince someone to let you plug in to an outlet in their garage and pick up the charge you need.

HE SAID:

My wife has a friend who is violently opposed to owning a plug-in vehicle. "What if the battery runs down? I can get gas on almost any corner but I don't know where there are chargers." Here's what she's missing.

You will do at least 80% of your charging at home, overnight. You may even do 100% of your charging at home, if you don't take road trips. Charging should really be a non-issue for most Model 3 owners. It is one of the most unappreciated gifts of owning an EV. You don't have to keep checking the fuel gauge to determine when you'll have to drop by a gas station to refuel. You have a *gas station* right at home and your car is all fueled up EVERY MORNING. How great is that?

There are a couple of simple rules you should follow when charging. Don't charge your battery to 100% except for the night before a long trip. Frequent charging to 100% will shorten the battery life significantly. Most Model 3 owners charge to 80% or 90% nightly. Elon once tweeted: "Not worth going below 80% imo. Even 90% is still fine. Also, no issue going to 5% or lower SoC."[9] Never let the battery run completely to 0% charge. This could damage components in the car and will require a tow to the nearest Tesla Service Center to get the car running again.

TWO BATTERIES

The Model 3 has two batteries: A standard 12-volt battery which powers almost all the accessories and the 350-volt battery (the main battery) that drives the motors and the heater/air conditioner in the car. The 12-volt battery's charge is maintained by the main battery. This is one reason why it's dangerous to let the main battery run down to 0%. If it cannot recharge the 12-volt battery and that battery dies, then you can't even put the car into **Tow Mode** to get it towed to a service center.

If this happens to you, DON'T PANIC. There's a solution and it's detailed in the Tesla Owner's Manual. Look in the section called *"Instructions for Transporters*" under *"If Model 3 Has No Power."* It's a little like getting a jump start. You'll see...or, hopefully NOT.

WHY IT'S GREAT...

Feeding an EV is cheap especially compared to filling a car with gas. I pay about seven cents per kilowatt hour (kWh) for electricity where I live. Our

Model 3 is a LR AWD with a 75kWh battery. That means we can charge her from 10% State of Charge (SoC) to 90% SoC (an 80% difference) for about $4.20 (80% x 75kWh) x $0.07). I can **fill my tank** for $4.20. They were charging more than that for a single gallon of gas the last time I was in California! Even if you live where electricity is more expensive, say Los Angeles, and you are in the most expensive pricing tier (21.6 cents/kWh as of this writing) the calculation works out to be just $12.96. Try getting a tank of gas anywhere for that little. And, if you have solar panels on your roof (as we do), you can drive on sunlight!

THREE BASIC OPTIONS FOR CHARGING AT HOME (in the USA)

GOOD: From a standard 110 VAC (NEMA 5-15) electrical outlet, using the Tesla adapter and cable. Not really practical. This will only add about 3 miles of range to your battery per hour of charging time.

BETTER: From a 240 VAC outlet (NEMA 14-50). The type of outlet required by an electric clothes dryer. This isn't a terrible option. With this hook-up you should be able to add about 30 miles of range to your battery per hour of charge time.

BEST: A Tesla Wall Connector wired to a 240 VAC 60AMP circuit breaker will afford you 44 miles of range for every hour of charge time. The price for the wall connector was about $500 (purchased from Tesla.com) and an electrician will cost at least a couple hundred more. We did it though and we're glad we did. Depending on your state you could qualify for rebate or tax incentives that will offset some or all of this cost.

CHECK OUT: Installation Manuals—Wall Connector
tesla.com/support/installation-manuals-wall-connector

CHECK OUT: Find an electrician via Tesla Support: Tesla.com/support/find-Electrician

HOME CHARGING WITH METERED RATES

If cost of electricity at your home is cheaper after certain hours (metered rates) you don't have to try to remember to plug-in then, you can set the charge menu on your Model 3 for the time you want it to start charging and get this: it remembers this setting and associates it with this location. So, if you charge only after a certain time at home but use an EV charger at work, you don't have to keep toggling the scheduled charge time. The Model 3 will charge normally while you're at work, or other location, but will remember to only charge after the scheduled time when you are home. How cool is that?

Simply plugging a charging cable into your Tesla is all it takes to initiate a charging session. The session ends which the battery achieves the SoC that you selected. To set a charging time, TAP the BOLT ICON on the DRIVING PANE to bring up the CHARGING SCREEN. At the bottom of this window will be charging options, including start time.

TESLA SUPERCHARGING

A ROAD TRIP is the second-most fun thing you can do in your Model 3. At the time of this writing, there are over 14,000 Tesla Superchargers worldwide. (Tesla claims 2019 is an expansion year, expecting to double this number by 2020.)

COST

At the time of this writing, Tesla's website[10] says the average price for Supercharging in the United States is $0.26 per kWh. Pricing varies by location, however, and at some locations pricing is per connected minute. Per-minute chargers use a tiered approach to pricing where Tier 2 applies to vehicles charging at over 60kW and Tier 1 (half the per minute price of Tier 2) applies to cars that are either sharing a Supercharger with another car (see Tips below) or charging at 60kW or less.

The Tesla Supercharger map at:
tesla.com/supercharger or
tesla.com/findus/supercharger

WHY IT'S GREAT...

- Tesla states on their web site: "Tesla is committed to ensuring that the Supercharger will never be a profit center." In other words, they will set charging fees only to offset the actual cost of power and maintenance of the network.
- They've also tried to locate the Superchargers near hotels, restaurants and shopping areas so you can grab a bite or do some light shopping while your car charges. Much better than trying to dine on gas station fare.

PRO TIPS:

- **SHARED CIRCUIT:** When possible, try not to share a circuit. Superchargers are generally labelled with numbers and letters like: 1A, 1B, 2A, 2B, 3A, 3B, etc. Superchargers with the same number share a circuit this means the second car to plug in will charge at a slower rate until charging for the first car has slowed or completed charging. Choosing a shared Supercharger circuit when others are open is considered bad Tesla etiquette.
- **IDLE FEES:** At a Supercharger, disconnect when charging is complete, even if it means moving your car, to avoid **Idle Time Fees**. Tesla does this to ensure that chargers are always available to drivers who need them. When at least half the chargers at a location are in use, idle time fees apply to anyone who does not disconnect within five minutes of charge completion. Don't worry. If you're in a restaurant, the Mobile App will keep you apprised of charging progress and tell you when it is complete.
- **BATTERY PRECONDITIONING:** In cold weather and certain other conditions, the battery will benefit from some **preconditioning** prior to charging. If you set your GPS to navigate to a Tesla Supercharger a message may appear on the Driving Pane of the

touchscreen that says: "Preconditioning battery for Supercharging." If you see this message, it's a good thing. If you don't see it, don't worry. It just means your battery is already warm and ready.

- **DESTINATION CHARGERS:** You plug in at night when you're home but what about staying in a hotel? Many of the larger hotels chains have added EV destination chargers.

✔ **CHECK OUT:** Maps and information about destination chargers:
tesla.com/destination-charging
chargehotels.com

- **OTHER CHARGING OPTIONS:** Currently, only Tesla vehicles can charge at Tesla Superchargers. Teslas are also able to charge on other charge networks though, such as, EVgo, ChargeHub, ChargePoint and Volta. To do this, you will need to use the J1772 adaptor that comes with your Model 3. We've found it's convenient to just keep that little accessory in the glovebox.
- It shouldn't happen, but if for some reason your charging cable will not release from the car, open the trunk and look inside the area directly behind the charging port. There is a manual emergency release system. Follow the instructions and be grateful that Tesla seems to have thought of everything. #ElonFTW!

VAMPIRE DRAIN

This is one of those Tesla topics that forum followers love to sink their teeth into. Your Model 3 will use some battery even while it's parked. Tesla owners refer to this as vampire drain. It's a product of the car trying to keep the main battery within a healthy temperature range, communicating with the Tesla network and various other sundry tasks that occupy its idle time. The manual says this drain is normally the equivalent of 1% of range per day with minor variations due to weather and other factors. This is clearly not an issue if your Model 3 is at home and plugged in but what if you've left her in an airport parking lot?

There's no magic solution here. But you should plan to park your car with sufficient charge so you will still be able to get home—or to a Supercharger—when you pick up your car upon returning from your trip. So, if the airport's 40 miles from your home and you plan to be away for ten days, make sure you park the car with more than 50 miles of range in the battery. Not that hard, right? If you're really worried about it there are some pro-active steps you can take when you park your car, like turning off Sentry Mode and Cabin Overheat Protection. Or set Cabin Overheat to fan only. Also, this might be the hardest one, but resist the temptation to check in on your car via the MOBILE APP. Every time you wake up your car you will sip a bit of charge.

BATTERY RANGE DEGRADATION

Don't let anyone scare you about overall battery life of a Model 3. Yes, you will lose range over the life of your battery. This is a fact for all Battery Electric Vehicles (BEV). The loss, however, is turning out to be less than originally anticipated. Tesla guarantees the battery to retain at least 70% of its original capacity up to 100,000 miles of driving (120,000 miles for Long Range battery). Real world experience is actually a brighter picture.

First, you should know that the loss in capacity is not linear. According to a source[11] we found, a survey of about 350 Tesla owners in Europe revealed a 5% drop in range by 50,000 miles. After that however, the curve flattens significantly showing that, on average cars driven to 160,000 miles still retained 90% of their original charge capacity. The authors of this survey believe their numbers suggest you may retain as much as 80% of capacity up to 500,000 miles.

BATTERIES AND YOUR CARBON FOOTPRINT

Lastly, there are always concerns about the disposal of these batteries. Some people will claim that *spent-batteries-filling-landfills* mean ICE vehicles are actually better for the environment.

This is a dubious claim to be sure, especially considering that Tesla's Lithium Ion batteries are recyclable. In fact, Tesla has stated in their latest impact report[12] that they are developing a unique battery recycling system at Gigafactory 1 that will process both scrap from the battery manufacturing process and end-of-life batteries to recover valuable materials, such as, lithium and cobalt as well as all metals used in the battery. These recovered materials will then be recycled into new batteries.

4

BASIC OPERATION

"I'm interested in things that change the world or that affect the future and wondrous, new technology where you see it, and you're like, 'Wow, how did that even happen? How is that possible?"

—Elon Musk

SHE SAID:

If you use a smartphone or tablet, operating the 15" touchscreen will be instinctive. If you're not that comfortable with the technology, don't let it intimidate you. Driving and customizing your Tesla is nowhere near as complicated as resetting the clock in basically every vehicle, stove and microwave I've ever owned. Relax, most of the important features are automatic.

Think of this section as a QUICK START menu. First things first.

DOOR (out → in)

PRESS your thumb into the large area of the handle. When the rest of the handle emerges from its recessed location, grasp and pull.

DRIVER PROFILE

(quick access) Reach down and use a lever to adjust the driver's seat. This action brings the **driver profile DIALOG BOX** onto the touchscreen. Select ADD and fill in your name. Included within this dialog box are ICONS for adjusting the settings of the rearview mirrors and the steering wheel. Select and use the LEFT scroll button on the steering wheel to adjust. (Scroll UP or DOWN and CLICK RIGHT or LEFT) Once the seat, steering wheel and mirrors are to your liking, select SAVE on the touchscreen to retain this setting. (See **Chapter 5: Bells & Whistles** for more preference settings.)

GEARS

RIGHT STALK: UP for **REVERSE**, DOWN for **DRIVE**. **NEUTRAL**, halfway UP or halfway DOWN and hold for a second or two. **PARK** PRESS IN on the button located on the tip of the RIGHT STALK.

START

Buckle up and PRESS the brake pedal, select a gear and GO!

DOOR (in → out)

Slip your hand (thumb up) into the door grip opening. Above your thumb locate a triangle-shaped button. This is the electronic door release. Press with your thumb to open the door.

From here Your Model 3 drives like a regular car.

WHERE TO LOOK ON THE TOUCHSCREEN

We're not including photos of the Model 3 touchscreen layout because browsing YouTube videos and photos of the screen from even only a few months ago we notice that the placement and number of icons on the screen has changed. We can be reasonably certain they'll change again. The important thing is to *learn the basics* and *remember where to find what you're looking for.*

TOUCHSCREEN BASICS

The touchscreen is divided into two main areas. The left, 1/3 of the screen—nearest the steering wheel—is what I call the DRIVING PANE. The 2/3 area to the right is the NAVIGATION PANE.

KEEP YOUR EYES ON THE DRIVING PANE

To monitor speed, gear, range, Autopilot availability/use, TACC availability/use, speed limit, plus a rather large, super-cool, real-time, animated AVATAR of your actual car while parked and/or driving, along with any vehicles or objects detected around it.

PRO TIP:

Nothing—no menus, warnings or other distracting information will ever obscure the upper quadrant of the DRIVING PANE. Even if you happen to accidentally tap the wrong button somewhere else, still nothing will cover THIS screen while either driving or parked. So, no worries. Tap away. (NOTE: fully opening the climate controls window can obscure the lower part of the DRIVING PANE, but not the area necessary for driving.)

AVATAR

The avatar so accurately represents your car that it even displays an open door, open trunk, if your headlights are on, as well as vehicles and other obstacles next to you on the road. The camera even recognizes pedestrians and bicyclists and shows their likeness on the screen.

WHAT'S ALL THAT BELOW THE AVATAR?

REAR BACKUP CAMERA accessible while parked and while driving.

Access charging menu

Microphone for Voice Activation: use for navigation, media and hands-free phone calls.

Manual windshield wiper controls if AUTOMATIC isn't cutting it. This is also where you will find the WIPER OFF setting for car washing. UNDER the WIPER ICON you'll notice three DOTS. Swipe the WIPER ICON to the RIGHT or LEFT to reveal the other menus indicated by the dots.

TRIP ODOMETER SWIPE RIGHT on WIPER ICON to find two "since last" odometers, two definable trip odometers and one regular odometer.

TIRE PRESSURE SWIPE LEFT on the WIPER ICON to reveal the tire pressure monitor. To display current PSI, you must be driving over 10 mph for fifteen minutes.

TAP the car ICON at the bottom of your screen to access the DRIVING PREFERENCES menu. This menu will open over the NAVIGATION PANE. (NOTE: This menu is where most of the customization takes place. You will want to know how to find this menu on your car.)

MUSIC & MEDIA APP - Self-explanatory

APP LAUNCHER: leads to ENERGY APP, WEB BROWSER and other Tesla Apps. The Energy App is important for monitoring range while on a trip.

THE KEY–3 OPTIONS

If you're one of those people who even take your smartphone with you to the bathroom, you're going to LOVE Option 1.

Option 1: Your smartphone authenticated to communicate with the car via Bluetooth.

Option 2: A sexy black, credit card-style key card.

Option 3: An optional (for a fee) key fob.

MOBILE APP: POWER IN YOUR POCKET

Works with iPhone or Android smartphones.

(Your Tesla delivery Rep will probably set this up for you, but you should still know how it's done.) What you need: Smartphone, Key Card and Tesla Model 3.

1. INSTALL Tesla Mobile App on smartphone and LOG IN using your Tesla ID and password.
2. Turn ON Bluetooth (phone) and ENABLE Mobile Access (app).
3. On app: TOUCH PHONE KEY then TOUCH START to search for Model 3.
4. When Model 3 is detected Mobile App will prompt to **TAP your Key Card**.
5. TAP (or touch) your Key Card against Model 3 card reader located on the center console, behind the cup holders. APP will confirm that phone has been authenticated. At that point PRESS DONE on the APP to save authentication. (If this fails, follow steps in **Chapter 8: The Geek Stuff** > Software Hangs, Lags & (maybe) Quick Fixes.)

WHY IT'S GREAT...

- Convenience. SHE SAID: Bliss = My life all in one place! (Note: the Mobile App also conveniently displays total miles driven and the VIN number.)
- When using the Mobile App as your car key, once you reach your destination, you simply put the Model 3 in PARK. Get out of the car and walk away. The Model 3 will detect your absence (whine softly) then turn off the car, flash the headlights and lock the doors. It will even (via a preference setting) issue a short chirp as audio confirmation. Boom! Done! (NOTE: AUTOLOCK ON EXIT is a configuration. It's likely that the dealer will set this up for you, but just in case, you can find the setting at: DRIVING PREFERENCES > LOCKS > Autolock (toggle on or off).
- The Mobile App controls other basic operations, too. From any location, via the Mobile App, you can: lock and unlock the vehicle, flash headlights, honk the horn, start the motor, open front or rear trunk. Set valet mode, sentry mode, as well as, set climate controls. If the car is ON a media control will appear on your Mobile App allowing you some limited controls over what is playing in the car. The Mobile App allows you to monitor charging status and engage Summon. (Seriously, if it did windows I'd marry it!)

- From the Mobile App you can see an exact mapped location of your Model 3, both when it's parked and if it's moving.

PRO TIPS:

- MOBILE APP + CLIMATE CONTROL: Check the temp inside your car. Then adjust it (warmer or cooler) before leaving your house or place of business. Voila! No more sweaty pits or frozen fingers. NOTE: if you park outside in freezing temps also select window defrost for clear sailing on your drive and to help unfreeze wipers.
- MOBILE APP + CALENDAR SYNC: On the Mobile App (gear wheel menu) enable CALENDAR SYNC. This syncs appointments and dates from your smartphone calendar to your car. If your calendar entry includes an address, tapping the entry on the touchscreen will automatically set this as your destination and navigation will begin.
- MOBILE APP + NAVIGATION: Look up an address on your phone in Google maps (while you are in the house or in a hotel room). SELECT SHARE to the Tesla Mobile App. This sends the destination to the Model 3 navigation which will be up and running when you enter the car. Extra points for NOT holding up a valet parking line while you TYPE IN or SPEAK your destination.
- CONTINUAL MOBILE APP CONNECTION PROBLEMS: Follow instructions in **Chapter 8: The Geek Stuff** > Software Hangs, Lags & (maybe) Quick Fixes.

KEY CARD OPERATION

Move over black AmEx, the Tesla Key Card is equally sexy. You will receive two of them with your Model 3. These cards are keyed and authenticated to only your vehicle.

- Operation is simple. There is a **RFID card reader** mid-way on the door pillar on the driver side of the car. TAP or TOUCH Key Card to that area to unlock the door. Once inside the vehicle, place the key card on the console behind the cup holders (location of the interior card reader) and leave it there for the duration of your drive. Be sure to take the card with you when you exit the vehicle. (IMPORTANT NOTE: the Key Card will **not** Autolock your car like the Mobile App. To lock your Model 3 with the Key Card, tap the card on the door card reader as you leave the vehicle.)

- The Key Card is your backup if you lose your phone or if the phone battery dies.
- Use Key Card for Valet Parking

WHY IT'S GREAT...

- You don't have to leave your phone in order to use valet service.
- Back-up systems are sane and extremely important when dealing with technology. The key card is your back up to the smartphone key. Keep at least one key card in your wallet. #backupFTW.

PRO TIP:

As Tesla Model 3s become more and more ubiquitous it's easy to imagine a valet having a handful of Key Cards to deal with. Order or make a unique-looking sleeve with your name and phone number on it to make for quick identification, and in case the valet has a problem with your car he can reach you.

KEY FOB

The key fob is available from Tesla for an extra fee. It can lock and unlock all doors, unlock the trunk or frunk and handle basic summon. Since you have to buy one through Tesla, anyway, see the Tesla Owner's Manual for how it works. Tesla charges extra for the FOB.

*At this time, it's $125, but the cost and operation are subject to change.

DOORS AND LOCKS

OPENING AND CLOSING:

The doors on the Model 3 are controlled electronically, which is a great feature for a number of reasons. It's why you can unlock the door to your car from anywhere. It's also why Tesla Roadside service can provide instant relief in a lock-out situation. And, it's why when you open any of the doors on your Tesla Model 3, the electronics also slightly lower the corresponding window so that the unframed window glass won't bang against the body of the car when you close the door. In the case of a no-power situation, Tesla built in an emergency release latch on the driver and front passenger door. The problem is the emergency latch is located where one would expect to find the door handle. So many passengers accidentally deploy the emergency latch.

Initially, at the time we bought our car, the emergency release latch did NOT lower the window so doing this was a problem. Thanks to a recent OTA update, however, the emergency latch now also lowers the window which means the glass isn't in any real danger. But you will likely get a message on the screen saying *"Manual door release used. May cause damage to window trim."* Don't panic! If your car is on software version 10 or higher your window is protected.

The electronic door release button is the triangle-shaped button located at the top of the door grip handle.

LOCKING:

There are no physical lock/unlock mechanisms on the inside of the Model 3 doors. Don't panic! To unlock the door to allow a passenger outside the vehicle to get in, the lock/unlock control is digital. It's the PADLOCK ICON located on the top of the touchscreen, just to the right of the DRIVING PANE.

Just like the padlock on your smartphone it is toggled ON or OFF via a TAP. You can also unlock all doors from outside the vehicle via the Mobile App, regardless of how far you happen to be from your car. (Example: you're out to dinner with a friend but your car is home and your daughter calls with a panicky message saying she left her homework in the backseat of your car and it's locked. You can, from a distant location, unlock your car allow her to retrieve said homework and then re-lock the car again.)

HAVE IT YOUR WAY:

There are options. When picking up passengers, come to a complete stop and put the vehicle in PARK by pressing the button on the end of the RIGHT STALK. Press this button a second time to unlock all the doors to allow a passenger to get in. Once you put the Model 3 into drive or reverse, the doors automatically lock again. This way the doors stay locked until you want them unlocked. Another option is to configure the vehicle to simply unlock all doors whenever you put the car into PARK. Go to the DRIVING PREFERENCES menu > LOCKS and set UNLOCK ON PARK to the ON position.

ACCIDENTALLY LOCKED OUT?

It seems like this would be nearly impossible given the key configurations for this car, but from what I've seen on Model 3 forums it happens more often than it should. Fortunately, the solution is way easier than being locked out of an ICE vehicle i.e. locksmith not required. In fact, there is even more than one option. If for some reason your smartphone, key card or key fob are not within reach you can call your partner and get him or her to use their smartphone to unlock the vehicle. Or in a pinch, you can call Tesla and THEY can unlock your car remotely. (Note: Tesla Roadside assistance is available 24/7/365.) How sweet is that? Technology FTW!

CHILD PROTECTION LOCK

Yes. The Model 3 has them. Access the DRIVING PREFERENCES menu > LOCKS and set the feature to the ON position. Please read the Tesla Owner's Manual for all safety information regarding car seats and child protection features.

CLIMATE CONTROLS

SHE SAID:

I run a little hotter than my husband and I need air blowing on my face when I drive. We've been married twenty years so obviously we've managed to make this work in all of our other cars but NOW that we have Pearl, I have no clue how we accomplished that. The Climate Controls in the Tesla Model 3 are a thing of beauty.

HE SAID:

She's always saying, "It's warm outside, why are you wearing a sweatshirt?" That's how we made that happen!

Check out that sleek, futuristic Model 3 dash again. There are exactly zero plastic vents and louvers to direct air flow around the cabin. And yet you can pinpoint exactly where you want the air directed and how much.

PRO TIPS:

- TAP the TEMP SETTING (lower menu bar) to adjust the temp up or down. You can either tap, tap, tap on the directional arrow OR utilize the slide bar that appears right above the temp on the screen.
- To adjust the air flow direction: TAP the FAN ICON. This opens a seriously cool 3D visualization of how the air circulates. It's completely controlled by a second blade of air. TOUCH, DRAG, PINCH or WIDEN that interactive airflow around on the touchscreen until the flow is to your liking. At this point you quickly realize you're not in car-Kansas anymore! I'm betting if you didn't see this before you bought the car, you'd buy it all over again...right on the spot.

  But wait, it gets better!
- Tapping the TEMPERATURE icon will reveal a sync button which splits the climate controls between the driver and passenger side of the car. This means I can have colder air blowing on my face and my better half, with sensitive dry eyes, can adjust his airflow differently.
- To get AC or HEAT into the BACKSEAT you need to turn it on. TAP the FAN ICON to bring up the CLIMATE CONTROL menu. The control for backseat climate is located on the lower, far-RIGHT side of the touchscreen. It looks like a double seat, front and back.

SHE SAID:

- If you're dealing with an obnoxious backseat driver, go ahead let 'em suffer...but if it's your kids or your besty, you'll want to know where to find this control.

CHECK OUT:

Tesla Owners Online:
YouTube (4:40)
MODEL 3 HVAC SYSTEM EXPLAINED

HEADLIGHTS

The Model 3 headlights are automatic. This means the headlights, taillights, side marker lights, parking lights and license plate lights automatically come on in low lighting conditions. If you want a different setting, for example, you're at a Drive-in movie or maybe you find yourself parked on a dark, vista with steamy, romantic intentions. Once you completely discount that this is how horror movies often start, you'll want to turn the headlights OFF while keeping the car ON. You can turn the headlights off From the DRIVING PREFERENCES menu. But the lights will always revert back to the AUTO setting on your next drive.[13]

For high beams, PRESS the left stalk AWAY from you and RELEASE. The HIGH BEAMS will remain on. To turn them off PRESS and RELEASE the left stalk AWAY again. To flash high beams, PULL the left stalk TOWARD you and RELEASE.

WHY IT'S GREAT...

Going from bright outside sunlight into my dark garage the headlights automatically come on, illuminating my way. So easy and so safe. No longer do the cardboard boxes and sports equipment fear my return.

HAVE IT YOUR WAY:

There are even more headlight choices on the DRIVING PREFERENCES menu > LIGHTS. Headlights can be configured to automatically switch to high beams and then automatically switch back to low beams when oncoming headlights are detected. You can also configure HEADLIGHTS AFTER EXIT, which means the headlights and the exterior lights will remain on for a short period after you park and exit the vehicle in order to light your way.

WINDSHIELD WIPERS:

These too come on automatically when water is detected on your windshield. This feature is currently listed as "in Beta" and there has been some grumbling on the forums, so be prepared to go manual in the event of a downpour. For comparison, we live in the Pacific Northwest and have found nothing to complain about with the windshield wipers.

MANUAL OPERATION: Here are your options: if you need a quick swish or two PRESS the button on the end of the LEFT STALK. PRESS it again for extra swishing.

The windshield wipers can also be set for continuous operation in four configurations: intermittent slow/fast or continuous slow/fast. The windshield wiper icon is located on the DRIVING PANE just below the avatar of the car. TAP to configure.

SPRAY WIPER FLUID: fully PRESS and HOLD the button at the end of the LEFT STALK to spray fluid onto your windshield. The wiper will come on and continue for two wipes AFTER you release the button.

OTHER DRIVING PREFERENCES AND FEATURES

Now that you're piloting your Model 3 around like a BOSS, it's time to explore the real power behind this car. For convenience, the following features are arranged alphabetically, not in order of importance or which features we love the most. To be honest, we love them ALL.

12V POWER SOCKET

There is one, located in the center console rear compartment.[14]

AIRBAGS

For Child seat placement and airbags, please read the Tesla Model 3 Owner's Manual. These safety features are too important to be left to our layman explanations. Go direct to the source for this info. NOTE: The passenger side front airbag is OFF unless weight is detected in that seat. (See Airbag OFF indicator at top right of touchscreen.) Depending on placement, that weight can be as little as 20 pounds. If you're the only one in the vehicle and you do NOT see the icon indicating that the passenger airbag is off, you might need to clean out your purse or briefcase.

BACKUP CAMERA

Selecting reverse opens the huge, amazing back-up camera which displays on the navigation pane. But if you just want to see what's back there without putting the car in reverse there is a back-up cam button located on the

Driving Pane just below the car avatar area and it works at any time, even while driving.

CABIN CAMERA

As of this date, there is one. A camera facing the interior of the car. The manual reports it's not currently active but might be used in the future. I can only imagine what weird awesomeness Elon is cooking up.

CABIN OVERHEAT PROTECTION

The Model 3 is very self-aware not only does she know *where* she is at all times but she can monitor and report *her condition*. Example: she is programmed to monitor temperature inside the vehicle, even when no one is inside, because overheating isn't good for the battery. In the event that she determines the cabin temperature is over 105 degrees, she will initiate a CABIN OVERHEAT intervention by turning on the A/C and FAN, or just the FAN. Like nearly everything else on a Model 3, Cabin Overheat Protection is optional and can be configured via the DRIVING PREFERENCES menu > SAFETY AND SECURITY > CABIN OVERHEAT PROTECTION. Choices include: OFF/disabled, NO A/C (fan only) or ON which uses A/C to cool the interior of the car.

WHY IT'S GREAT...

Providing you have enough charge on your battery, Cabin Overheat Protection can run for a period of up to twelve hours. (IMPORTANT NOTE: if your battery level falls below 20%, the Model 3 *protects the core* by canceling Cabin Overheat Protection.)

DRIVER PROFILE

The Model 3 Driver Profile feature is extensive. Saved seat, side mirror and steering wheel settings are common configurations for this class of automobile. But for the Model 3 they're only the beginning. Your personal climate control settings are stored with your driver profile. Even the more complex Autosteer and Autopilot preferences, that are really the Wow-factor in the Model 3 experience, are saved and maintained there as well. Setting and saving your driver profile means you literally only need to TAP one-button and adjust the rearview mirror and you're good to go.

Because the Mobile App links your smartphone to your car, Driver Profile settings also include pairing with your phone for voice calls, music apps and calendar sync. Recently added (via OTA) was the ability for Model 3 to determine which mobile device enters the driver's seat and automatically selects that person's driver profile when the car is started.

Setting your driver profile will be the first thing you will want to do and your delivery person may even walk you through this at the time of delivery. Adjusting the driver's seat will automatically bring up the DRIVER PROFILE dialog box giving you the opportunity to create a driver profile, or adjust and save new settings to an existing profile.

WHY IT'S GREAT...

Who doesn't want to just get in their car, press ONE button and go? Do I have to remind anyone who perhaps owned a Prius that you practically had to offer up a sacrifice to get the car to accept pairing when switching from one phone to another? Hmmm...anyone?

PRO TIP:

The Model 3 will allow up to 12 driver profiles. Because a profile contains so many aspects of your driving preferences you can save more than one profile depending on the particular driving conditions. Example: chill mode + standard steering for low traffic Autopilot situations verses the more aggressive Mad Max lane change Autopilot for when you're in a time crunch.

DIVA PRO TIP:
SHE SAID:

Shoes! Okay, you with me? Flipflops require one seat setting and heels or platforms another. Simply set up more than one driver profile to accommodate your shoes and you will always be stylishly ready to roll at a single touch.

FOLLOWING DISTANCE

At the moment, all Tesla Model 3s come standard with Autopilot which endows your car with the ability to steer, accelerate and brake automatically[15]. Not only is this feature cool AF, but you can fine-tune it for your specific comfort level. For example, how closely do you want to follow the car ahead of you? The default setting is 3. This does not mean car lengths but is described as "the time-based distance representing how long it takes for Model 3, from its current location, to reach the location of the rear bumper of the vehicle ahead of you."[16] This distance is adjustable from one to seven.

To change the distance, while driving in TACC, use the RIGHT SCROLL BUTTON on the steering wheel. CLICK RIGHT to decrease the distance or LEFT to increase following distance. This distance can also be selected on the DRIVING PREFERENCES menu > Autopilot > Cruise Following Distance.

GLASS ROOF

A glass roof is completely different from a GLASS CEILING. To be sure, the glass roof is one of Pearl's more gorgeous and elegant features. But, like ALL of the strong amazing women I know, that glass roof is not just appearance without substance it is an integral part of why this car received a 5-star safety rating. (See **Chapter 7: Safety**.)

When our glass roof gets wet it turns a beautiful rusty, orange color and the droplets glow almost red. What causes of this color change is a patented, laminated reflective layer inside the glass that filters out UV and infrared light which contribute to heat (aka greenhouse effect). However, recent information implies that while the laminated reflective layers remain, the roof no longer turns red when wet. We live in the Pacific Northwest and the mild tint is enough even though the *Giant Ball of Death* (aka the SUN) gets pretty intense up here. But I have seen comments from people who live in Arizona and Southern California who say the interior of the vehicle still gets pretty hot.

SOLUTION: There are several aftermarket sunshades available.

HOMELINK (optional)

If your vehicle includes the HomeLink option, you'll be able to operate up to three Radio Frequency (RF) devices from your Model 3. The pairing is explicit and device-dependent, so we'll refer you to the Tesla Owner's Manual for detailed instructions. The manual offers a good amount of information on how to pair and also how to troubleshoot HomeLink problems. The manual even includes a phone number to HomeLink advising you to reach out to the manufacturer for compatibility questions and additional assistance.

WHY IT'S GREAT...

Our HomeLink settings work perfectly with maybe only the occasional hang up or inaction. And by perfectly, I mean that HomeLink opens and closes the garage door without us having to touch anything in the car or on the screen—except for the handful of times when the door hangs. Then all we have to do is touch the HomeLink icon on top right side of the touchscreen.

HE SAID:

This hang situation is probably just a glitch with our (rather old) door opener. It seems to only happen when I'm driving and, if I thought Sheryl knew enough about electronics to monkey with it, I'd be sure she was punking me.

SHE SAID:

You'd think after all these years he'd realize I have better ways of punking him!

Our HomeLink is set to close the door when it detects Pearl leaving, once she pulls into street or to open it when it detects her arriving at the end of the driveway. In typical Tesla HAVE IT YOUR WAY fashion you can configure the distance from the door or gate where you would like the activation to begin.

I've seen a few forum posts where HomeLink has misfired and brought the garage door down on a car. Not sure if the fault lies with owner error, HomeLink or Tesla. We haven't experienced this problem because the switch that opens and closes our garage door is on the wall, right inside the door from the house. We manually open the garage bay as we enter the garage and manually close it after we've parked. We do it this way for convenience because it also turns on/off the light. But this way also means our door will never misfire on our car.

PARK

To put your car in park, with your foot on the BRAKE PRESS IN on the button located on the end of the right stalk. (NOTE: the Model 3 automatically shifts into PARK if the driver opens the door to exit the vehicle.) To activate the **PARKING BRAKE**: PRESS in on the Park button and HOLD for a second or two. Putting the car in gear automatically removes the brake.

SCROLL BUTTONS

The scroll buttons on the steering wheel are two of my favorite features on the Model 3. Depending on what Mode you are in, they manage different things. The scroll buttons have FIVE basic functions: scrolling UP or DOWN, clicking RIGHT or LEFT and pressing IN.

MIRRORS AND STEERING WHEEL SETTINGS: When parked, the LEFT scroll button adjusts steering wheel and side mirror settings when the DRIVER PROFILE dialog box is displayed.

AUDIO VOLUME: LEFT scroll button adjusts the AUDIO VOLUME up or down. Clicking right or left will advance or reverse tracks.

VOICE ACTIVATION: PRESS in RIGHT scroll button when either parked or driving to access the Model 3 microphone. Voice activation can make a hands-free phone call, set a navigation location, request music or media to play and even file a BUG REPORT with Tesla.

SPEED CONTROL: When driving and using TACC or AUTOPILOT the RIGHT SCROLL BUTTON rapidly increases or decreases your speed without taking the Model 3 out of TACC or AP. Super convenient!

HE SAID:

She didn't even mention that the scroll buttons are used while playing some of the games available in the Toy Box. It's like the BEST use of those buttons!! Pew, pew, pew!

SHE SAID:

Sigh! My entire family games and I overlooked the game aspect of the scroll buttons on MY car. Accidental? Probably not!

SOUND SYSTEM

Operating the sound system in the Model 3 is not mysterious. The controls function like every other car only EASIER. In fact, iPhone to Model 3 Bluetooth pairing is the easiest experience I've ever had. If you're the only one who drives your car, this won't matter as much. You'll set your connection up once as part of your driver profile and won't have to do it again. But in our house, even for short trips, we take turns. One drives and the other programs the entertainment. TAP the MUSIC ICON to access CONNECT PHONE controls.

WHY THIS IS GREAT...

The sound system is incredible. We couldn't find a detailed description in the Tesla Owner's Manual of exactly how the incredible sound is achieved. But according to a Teslarati.com article[17] the Model 3 premium package sound system boasts a total of 15 speakers: 3 front tweeters, 3 front mids, 2 front woofers, 2 front immersive speakers, 4 full-range rear speakers, and 1 subwoofer. Thanks Elon, for such attention to detail. Also, the media can be separated from the other driver profiles so I can play music from my phone without affecting my husband's seat, mirror and climate settings. It should be noted AND appreciated that while the LEFT SCROLL WHEEL gives the driver control over the volume, track and pause. There is a SPEAKER ICON located on the bottom, far right of the touchscreen, within reach of the passenger, granting them this power, too.

PRO TIPS:

- A recent OTA update placed basic MEDIA controls on the MOBILE APP, so a passenger with a connected phone can control Media while it is playing in the car.
- The Model 3 comes with a one-year subscription to Slacker Radio. I'm not really a fan and instead stream music and other entertainment from my phone. As long as it's the connected phone it works seamlessly.
- Spotify access was added, so if you're a Spotify user it's easy to plug in your account. Now you can stream Spotify without your phone.

TRUNK, FRUNK & GLOVEBOX

The trunk holds more than I thought it would. The frunk is incredible bonus space. And the glovebox seems standard-sized. I love that the glovebox is locked and can ONLY be opened by a button on the CONTROL touchscreen. This is an added bit of security that I appreciate. You can unlock the trunk from outside the car if your smartphone is detected, as well as from the Mobile App. There is no exterior latch for the frunk. It opens either from the touchscreen or the Mobile App. But the glovebox can only be opened from inside the car. To access the glovebox, open the DRIVER PREFERENCES menu and look to the bottom.

PRO TIP:

Don't let the kids store their homework in the glovebox, the only access is from inside the vehicle.

USB PORTS (premium package)

There are two in the front, located underneath the phone dock shelf. These ports can be used to charge a phone (or two) and/or play audio files from a storage device. In order to use the Dashcam, a flash drive or other storage device **must** be installed into one of these front ports. (See **Chapter 5: Bells & Whistles** > **DASHCAM**.)

Additionally, there are two USB ports located on the back of the center console so passengers in the back seat have charging capability. Because heaven forbid any of the myriad of electronic devices that ride in this car should not have power **AT. ALL. TIMES!**

VALET MODE

Feel a little queasy about handing the keys to your baby over to a stranger? Me too! But sometimes there simply are no options, like when you're staying in a hotel or at a particularly crowded event. Valet Mode should give you a little peace of mind. Valet Mode will limit the speed and power acceleration, lock the frunk and glovebox, obscure home and work locations from the navigation system, as well as disable voice commands. Autopilot, Mobile Access, HomeLink and driver profiles aren't assessible either. Even Wi-Fi and Bluetooth are disabled and you can't pair a new device or delete any existing devices. ACCESS VALET MODE from the DRIVER PROFILES ICON at the top of the touchscreen. One touch puts it into VALET MODE, don't forget to hand them the KEY CARD.

PRO TIP:

You will need a four-digit PIN to enter and cancel Valet Mode. Do yourself a favor and set that up before you are holding up a long line of cars at your next Red-Carpet event.

VEHICLE HOLD

The Model 3 can continue to apply the brakes even after you have removed your foot from the brake pedal, like when you're parked, or even simply stopped, on a hill and don't want to roll into the car behind you. To engage HOLD, after coming to a complete stop, apply a little extra pressure to the brake pedal again. Watch the top of the touchscreen in the DRIVING PANE area. The indicator will display an H as the HOLD icon. To disengage HOLD, press the accelerator or pump the brake.

VOICE COMMANDS

The Model 3 accepts voice commands for Navigation destinations, media requests and for placing hands-free phone calls. To access the microphone, PRESS IN on RIGHT SCROLL WHEEL or locate the MICROPHONE ICON on the DRIVING PANE, below the avatar of your car.

WARNINGS

The Model 3 is a super-smart car. It's hard to even fathom the attention to detail that is contained in this car. Because it is so different from every other car I've ever owned or on the market today there is much to learn. Fortunately, the Model 3 is extremely patient. Through various warnings and alerts, this car will give you all kinds of information. One-line warnings display on the DRIVING PANE. The kinds of alerts you can expect are anything from "Keep your hands on the wheel" when employing Autopilot to "Press gently to close" if you're having trouble getting the plastic console cover to stayed closed. (*Go ahead and try it. I did after reading about a journalist's amazement. Seriously, how does it know? I don't know. But it does.*)

In addition to actual messages, there are various warning tones. Some tones are gentle reminder chimes, while others are screechy, *get-your-shit-together-man* warnings. Some tones are for assistance like PARKING SENSORS to keep you from banging into things. Others, like LANE ASSIST and COLLISION AVOIDANCE are serious safety devices engineered to save lives. But here's the thing, MOST of these warnings generate from configuration preferences found under the DRIVING PREFERENCES menu. For each individual driver you can select what level (if any) of warnings you prefer. (See *Chapter 7: SAFETY* for more options.)

5

BELLS AND WHISTLES

"I think the best way to attract venture capital is to try and come up with a demonstration of whatever product or service it is and ideally take that as far as you can. Just see if you can sell that to real customers and start generating some momentum. The further along you can get with that, the more likely you are to get funding."
– Elon Musk

HE SAID:

Tesla employs some really smart software developers. As a former software engineer, myself I can tell you that we are never really finished with any project. There's always one more feature, one more slight improvement we want to squeeze in there. The people at Tesla are no exception. Model 3 is equipped with a number of really nice features and they're adding new ones and making improvements all the time. The beauty of this is that these features and bug fixes come to you automatically Over-The-Air (OTA).

OVER THE AIR UPDATES

I've never before owned a car that constantly received new features or improvements. If I had, I'm sure I would have expected to drive back to the dealership to get these improvements and probably to be charged for them. Thanks to OTA updates, the Model 3 we bought in January of 2019 will eventually have all the features of one built in 2020 or 2021.

The only exception to this is the Full Self Driving (FSD) feature, which will require a whole new computer board and the new Tesla custom neural network processing chips. Since we paid in advance for the FSD feature,

Tesla will install this hardware without additional charge when it becomes available. Beginning sometime in April 2019 all new Model 3s included the upgraded board and chips when they are built.

Here are some more of the software features in the Model 3, many of which were added via OTA updates.

DOG MODE

When exiting the vehicle, you can select DOG MODE which allows you to keep the climate control on so the cabin of your Model 3 remains cool and your pet stays safe and comfortable while you dash into the store. It also puts a BIG message on the touchscreen stating, "My owner will be back soon" and displaying the temperature in the cabin, so well-meaning passers-by don't smash your window to *rescue* your Tesla-spoiled pooch. (NOTE: Check the local laws in your area regarding leaving a pet in the car. It might be illegal regardless of the temperature settings.) To ACCESS DOG MODE: TAP the FAN ICON to bring up the climate menu.

DASHCAM

Have you ever just gotten lost online watching dashcam footage of one car accident after another in other countries like Russia or China? I have, and not only is it entertaining it's a great way to blow a couple of hours. It's like every car over there has a dashboard camera and they're not afraid to post the footage online. You can see babies tumbling out of minivans into the middle of the street (unhurt!). In fact, you can see several different videos of the same baby tumbling out of a minivan. You'll see cars banging into each other, people driving on the sidewalk or the wrong side of the road for no reason at all except nobody stopped them. It's wild!

With a Model 3 you will no longer have to watch the madness from the sidelines, you can be part of it. All you need is a properly formatted STORAGE DRIVE plugged into one of the front USB ports and then you need to know how to save and download those videos.

FORMAT THE DRIVE

You'll need a LARGE flash drive, say 64GB or larger. I'll explain why in a minute. Format the drive as "FAT32" if you're doing it on a windows PC or "MS-DOS (FAT)" if you use a Mac. Just one more step—and nothing works without this one—create a folder on the drive named: TeslaCam. That's it. Don't put anything else on the drive. Plug it into one of the two USB ports inside the front console (the rear USB ports are strictly charging ports for phones, etc.) If you've done all this correctly, at the top right of the touchscreen you should see a small camera icon containing a red dot. The red dot indicates the Dashcam is enabled for recording.

There are three types of recordings stored on the flash drive. First there is a one-hour loop recording everything from both the forward-facing camera, the rear-facing ones on the sides of the car and the rear backup camera. These are stored as one-minute clips, the new ones overwriting the oldest automatically to form the one-hour loop. These files occupy about 8GB on the flash drive and are found in a folder called RECENTCLIPS.

SAVING DASHCAM FOOTAGE

A TAP on the camera icon instructs it to save the last 10 minutes of video, those clips are written under a folder named SAVEDCLIPS which requires another 1.2GB of space each. When the save operation is complete you will see a green checkmark in the camera icon for a second or so before it reverts back to the red dot that tells you it's recording. Data in the SAVEDCLIPS folder is never overwritten.

In addition to these recordings, Sentry Mode (if enabled) will also write 10-minutes of video clips to a folder named SENTRYCLIPS for each Sentry event which will require another 1.2GB of space. (NOTE: you will have a lot of Sentry Mode events that aren't vandals but admirers checking out your car. In fact, there's a whole slew of Tesla video cam admirers on YouTube.)

As you can imagine, with all of these videos, eventually the drive will fill up. You'll know when this happens if you see the camera icon sporting an 'X' instead of a red dot. At this point, you must remove the drive, erase all the files (or copy and save them on another drive) leaving the TeslaCam folder intact. Reinsert the drive back into the USB port.

REMOVING DASHCAM STORAGE DEVICE

On the touchscreen PRESS and HOLD the camera ICON to PAUSE recording before you remove the flash drive. If it's paused the camera icon will show a gray dot in the upper right corner rather than a red dot. If you don't do this, you may lose some footage. When you return the drive to your car you might have to tap the ICON again to un-pause it.

NOTE: Some folks on the Tesla forums have begun to suggest using an Endurance Micro SD card and USB adapter in place of the flash drive. They feel these cards will last longer and tolerate temperature extremes better. This issued is far from settled as other forum members say this claim is incorrect and using a flash drive is just fine if you choose a high-quality drive.

HE SAID:

We have two iDiskk 64GB flash drives. One is plugged in and active, the other stored in the console tray so when the active one gets full, I can swap them until I have an opportunity to clear the files off the full one.

PRO TIP:

Select a flash drive with a standard USB connector on one side and a Lightning connector on the other. I keep the iDiskk app on my iPhone so anytime I need to I can remove the flash drive and view the contents right there, on the spot, on my phone—super useful for showing the police that an accident wasn't your fault.

Not technical or don't have a computer? **Puretesla.com** sells Tesla formatted flash drives

SENTRY MODE

Tesla owners in the San Francisco Bay area were plagued by thieves smashing the small rear triangular windows on Model 3 vehicles which allowed them to reach in, flip the rear seat down and check the trunk for valuables.

Owners appealed to Elon Musk (via twitter) to use the side cameras on the car to capture images of these thieves for further police investigation. Within weeks, Tesla issued an OTA update containing the new feature called SENTRY MODE.

If you have a properly formatted storage device plugged into the USB port under the front console (the same device that supports your dashcam) and have Sentry Mode enabled, it will monitor the car's surroundings for activity. When motion is detected near the car, Sentry Mode goes into "Alert State" and puts a large message on the touchscreen indicating that the cameras are recording.

If the car is jostled (say, by another car hitting it in a parking garage) or if the window is broken, it goes into **Alarm State** which sounds an alarm[18] inside the Model 3 and will also alert your Mobile App that the alarm has been triggered. Video recordings of activity around the car beginning ten minutes before an alert or alarm event are stored on the flash drive as documentation for the owner and police.

So, you don't lose any footage PRESS and HOLD camera icon to PAUSE recording prior to removing the USB flash drive. This is similar to *ejecting* a flash drive on a PC or MAC and accomplishes the same thing. It makes sure any unsaved buffers are written to the drive.

IMPORTANT NOTE: If your alarm is triggered and wailing, UNLOCK the door to STOP the sound. This works either at the car with phone/key card OR via the UNLOCK icon on the MOBILE APP.

TRAFFIC-AWARE CRUISE CONTROL (TACC)

This may be the best thing to happen to your morning commute since drive-thru Starbucks. And, until we get the drive-thru Margarita bar it will elevate your evening commute as well. TACC frees you from that gas, brake, gas, brake, gas, brake dance we all do in stop and go traffic. PRESS the RIGHT STALK DOWN once and release it. The cruise speed will be set to the detected speed limit (adjusted for any offset you've specified in SPEED ASSIST) or your current speed, whichever is greater. The set speed is displayed inside a **BLUE** circle just below your actual speed on the DRIVING PANE area of the touchscreen. (If this circle is gray, it means TACC is not engaged.)

You can further adjust this set speed anytime you want by using the RIGHT SCROLL BUTTON on the steering wheel, or the plus or minus buttons next to the TACC speed. So long as there is no car ahead of you driving slower, your Model 3 will maintain this set speed. If the speed limit changes, your set speed **may not automatically adjust**. You can manually adjust it using the right scroll button or, the quickest way is to simply TAP the SPEED LIMIT ICON on the right side of the DRIVING PANE. This will adjust your cruise speed to the value displayed there.

The traffic-aware part comes into play when you have a car in front of you. If that car is driving slower than your set speed, your Model 3 will slow down to maintain a distance corresponding to the FOLLOWING DISTANCE specified in your Autopilot settings.

If the car in front of you stops, you will stop. When it starts moving again, you will start moving. Other than steering to stay in your lane you're removed from the stop and go game and can relax without worrying about losing focus and accidentally slamming into the car in front of you. This feature is intended for limited access highways, such as, freeways and turnpikes but it can even be used on city streets as long as you remember that it WILL NOT recognize or respond to stop signs or traffic lights. So, if there's no car in front of you or if he sneaks through as the last car on a yellow light, YOU must stop your car (at least until the Full Self-Driving feature is rolled out completely).

PRO TIP:

You don't have to be driving fast, on a highway or a long-distance haul to benefit from TACC. Example: the road through my town that leads to the freeway has a speed limit of only 25 mph. It's not a long distance, but it passes the police station. There's almost never any traffic and it's really hard to maintain a speed of 25 mph...especially in the Model 3. Engage TACC and TAP the speed limit icon. This sets your speed to the limit—25 mph—and you can cruise along without having to govern yourself. It's sublime.

AUTOPILOT (AP)

Autopilot refers to the combination of **Traffic-Aware Cruise Control** and **Autosteer**. At the time of this writing, AP is included in the base price of the

Model 3. You can activate TACC by itself but activating Autosteer always activates TACC. To activate AP, PRESS the RIGHT STALK down twice in rapid succession. Twin BLUE lines will appear on your screen from your AVATAR indicating AP is engaged.

AP uses a combination of eight cameras, twelve ultrasonic sensors and a forward-facing radar to keep an eye on the traffic around you and responds accordingly. It keeps your Model 3 it in the middle of its own lane and a safe distance from any car you are following. Of course, you can ALWAYS take back total control of the car by simply touching the brakes or maneuvering the steering wheel. Pressing the right stalk UP while driving will release AP and/or TACC as well.

When AP is engaged you still have to keep your hands on the wheel (or, at least nudge it occasionally when the car issues a warning). You must stay alert to your surroundings, but the car will handle all the curves and bends in the road on its own. The car will also maintain your speed, within your set limits, slowing—even to a stop, if necessary—in heavy or merging traffic without you having to press the brake or the accelerator.

PRO TIP:

There are many ways to configure the AP suite of self-driving abilities, depending on your preferences via the DRIVING PREFERENCES menu > AUTOPILOT.

WHY IT'S GREAT...

We've read that the radar on the Model 3 actually monitors the car ahead of the car ahead of yours. That's right, two cars up. Drivers using AP report incidents where their car initiated braking for no apparent reason...then, a beat later, the car ahead of them braked. Autopilot registered the need to slow and initiated braking before either driver had time to react.

AUTO LANE CHANGE

Ever heard "stay in your lane?" Yeah, screw 'em! If you purchased FSD, this function is available as soon as you engage AP. It allows you to request a lane change by simply turning on the turn signal. The car will speed up or slow down as necessary to safely insert itself into the indicated lane. Once it has successfully done so, it will turn off the signal automatically. This is good because I'm sometimes that guy driving down the freeway with his signal on for no apparent reason.

NAVIGATE ON AUTOPILOT (also part of the FSD package)

When you set a destination in the **navigation field** of the touchscreen and then activate AP, Model 3 will invoke Navigate-On-Autopilot (NOA)[19]. Now things get REALLY interesting. Your car will employ the features of AP and FSD with the purpose of reaching your destination. Depending on the options you configure it can automatically change lanes to pass slower traffic, negotiate any required route changes from one freeway to another on its own and exit the freeway at the appropriate off-ramp before turning control back over to you for the final leg of your journey on city streets or country roads.

Most drivers seem to feel this is the safest and best way to drive the Model 3, especially on road trips. Because you don't have to sweat missing the interchange or getting into the correct lane to make that switch, etc., you can focus on the traffic around you.

Of course, if you don't fancy using NoA, you can always turn it off by simply tapping on the blue Navigate On Autopilot button on the NAV route display. And if you don't want NoA to be the default mode when you enter a destination, you can change this by going to **Controls -> Autopilot -> Navigate on Autopilot**, which will disable it completely.

WHY IT'S GREAT...

You can configure how aggressive NoA is about getting you to your destination on time by selecting an option under DRIVING PREFERENCES menu > Autopilot > Navigate on Autopilot > CUSTOMIZE NAVIGATE ON AUTOPILOT > Speed Based Lane Changes. Your choices are DISABLED, MILD, AVERAGE and MAD MAX. Seriously—MAD MAX! I don't know if you are required to spray your teeth "all shiny and chrome" to use that but...maybe.

PRO TIP:

The Model 3 will learn and remember your home and work locations. You clearly do not need navigation to get to either one. BUT if you are using AP and you put the destination in your NAV, you will benefit from NOA, which will traverse all freeway transitions and handle lane changes from on ramp to off ramp. To quickly engage NOA to a home or work location simply SWIPE DOWN on the NAV dialog box. Based on time of day and your location, the Model 3 will likely know exactly where you want to go. One more thing, unlike the NAV in my other cars, I can set this one to home while driving.

NAVIGATION ADDITIONS

"What's it gonna be, punk? ... Do you feel lucky?" ...or hungry? Either way your Model 3 has you covered with the addition of LUCKY or HUNGRY buttons to navigation. Lucky takes you on a surprise adventure to a nearby attraction and hungry will find and display nearby restaurants.

AUTOPARK

SHE SAID:

A perfect parking spot is the unicorn of city driving, which is why I generally employ NORM, my parking Genie, to find me a proper space. Norm is extremely reliable. But once the space is located, I have to maneuver the car into it by myself.

The Model 3 has a much beefier parking Genie which they call **AUTOPARK** (part of the optional FSD package). I honestly didn't think this feature worked all that well until my husband demonstrated it for me. NOW it's one of my favs.

Autopark can parallel park or park in a perpendicular space. This feature made me absolutely giddy because you need to back in to most of the spaces at Tesla Superchargers and backing up in one straight, smooth motion is admittedly not one of my super-powers.

To auto-parallel park cruise up to a space just as if you were going to park yourself. When you are properly aligned a **P** will appear on the DRIVING PANE, to the right of your avatar. This means the Model 3 has detected a suitable space and is ready to take control. Release the steering wheel, shift the Model 3 into reverse, then touch START AUTOPARK (on the DRIVING PANE). The car will do the rest.

To Autopark in a perpendicular space there must be a vehicle parked on each side of the empty space. Pull slowly into position just as you would if you were going to back into the space on your own. Once you see the P in the Driving Pane, remove hands from steering wheel, put lever into REVERSE and touch START AUTOPARK. Then sit back and be amazed!

PRO TIP:

Touching or interfering with the steering wheel while the car is auto-parking will cancel the operation. Most of the time you will have the ability to select RESUME. But the manual very clearly states to stay aware and ready to apply brakes or intervene to avoid obstacles or pedestrians when using this feature.

SUMMON

If you live, work or travel in a city it's bound to come up eventually, that moment when you return to your car to find that some jerk has parked so close to the driver side door that you either need a can opener to get in or you're going to have to get in on the passenger side and climb over the console to get your car out. Tesla has an answer for that, providing you've purchased FSD. It's called Summon. At its simplest, it allows you to stand outside your car and use the Mobile App to drive your car straight out (either forward or backward) from the space you are in. The maximum distance the Model 3 will travel this way is currently about 39 feet.

If you're a bit more adventurous, there is **Smart Summon** which will "unpark" your Model 3 in a parking lot then drive it right to you, avoiding other cars, pedestrians and (most) obstacles. That's right. You will be able to HAIL your own car. How sweet is that?

IMPORTANT NOTE: Both Summon and Smart Summon are considered BETA. Tesla wants you to know that you are responsible for your car, just as if you were behind the wheel. You should have it in sight at all times and be wary of fast-moving pedestrians, bicycles and cars because it may not detect all obstacles.

FULL SELF-DRIVING (Optional upgrade)

If you purchase the Full Self-Driving (FSD) feature, you will get Navigate on Autopilot, Auto Lane Change, Autopark, Summon, Smart Summon and – eventually - the ability for your Model 3 to self-drive on city streets. FSD can be added via an OTA update after you purchase your car but it may be more expensive than including it in the original vehicle order. Check with your Tesla salesperson to be sure.

6

SERVICE AND MAINTENANCE

"I don't spend my time pontificating about high-concept things; I spend my time solving engineering and manufacturing problems."

—Elon Musk

TIRES

By now, you shouldn't be surprised to learn that the tires for Model 3 were specifically designed for THIS CAR. These tires have special treads to handle the weight (4,000 lbs.) and the instant torque generated by its electric motors. They are also acoustically insulated to reduce the amount of road noise that makes it into the cabin of the car. That's right. There's a strip of foam the width of the tire and about three-quarters of an inch thick glued to the inside of the tire to help absorb some of that annoying rumble that comes up from rough pavement. Who else but Tesla goes to that much trouble to make your ride super-comfortable and QUIET?

The Model 3 doesn't come with a spare tire. Does that seem weird to you? It seemed weird to me until I checked around and found out that about one third of all new cars today don't come with a spare[20]. It's probably just as well because given the weight you have to be super careful about jacking it up and the battery is pretty much the entire bottom of the car. You don't want to punch a hole in that. Not even a small one.

If you get a flat tire on the road Tesla does not recommend changing it on site, in which case you might need to be TOWED. (See the section on

TOWING and/or the Tesla Owner's Manual on **Transporting the Model 3** for complete instructions.)

PRO TIP:

Check with your Tesla Service Center. Currently in our area, in the event of a flat tire, we have been instructed to call Tesla Roadside Assistance (available 365/24/7). A Tesla Ranger will deliver a loaner tire to the car, put it on and take the flat tire back to the service center for repair or replacement. Then they will bring the new or repaired tire back to our location and install it. This level of service may vary across the country so it is best to check in advance. Also, a TIRE REPAIR KIT is available at Tesla.com which can offer a temporary solution to a flat.

JACK LIFT PADS

If you are going to rotate your own tires – or even take it to your local tire shop for that service—you should purchase a set of JACK LIFT PADS (they look like hockey pucks with a knob on one side. (**HE SAID**: *I wanted to call them "nipples" but my wife says "knob" is friendlier. Okay, knob it is.*) Make sure whoever is jacking up the car knows how and where to use them. Otherwise, your car could get jacked up...and not in a good way.

So, how do I use them, Jerry? Well, just like most other cars there are JACK POINTS identified on the bottom of the Model 3. Unlike most cars, these jack points have a hole in the middle of them. The knob on the top of the jack lift pad fits right into that hole which keeps the pad centered over the jack point. Most jack lift pads also have a rubber gasket around this knob so it will stay in place while you position the jack directly beneath it. Using these pads will create some distance between the edges of the jack and the body paneling or the battery casing that surrounds the jack point. If you take your car to a tire shop, be sure to give them all four pads and explain where and how they should be used, since they will probably want to lift the entire car on a hydraulic lift.

My manual says to rotate the tires about every 10,000 to 12,000 miles or, when the tread depth difference is 2/32 of an inch or greater. When it is time to rotate the tires it may be easier to just have Tesla do it. In most cases, they will send a Tesla Ranger to your home or place of work and do it right there. Cost is about $70. Certainly, more than your tire store would charge but you'll know it was done right. We set up a tire rotation at 10,000 miles, but when the Tesla Ranger checked there was virtually no difference in tread depth all the way around the car so, no rotation...and no charge.

Of course, you can rotate your own tires if you have a decent hydraulic jack and a heavy-duty jack stand. If you decide to do this, there is one more thing you will want to be cautious about. The rims on the Model 3 wheels protrude BEYOND the sidewalls of the tires. That means, if you let a wheel fall on its side while handling it, it will likely scrape some paint off the edge of the rim. So, you know, don't do that.

AERO WHEEL COVERS

A word about the standard issue Aero wheel covers. They're not the most eye-popping wheel configuration, but they *are* substance over pretense. Before ditching them just know, according to Tesla, they are designed to extend the range of your vehicle.[21]

TOWING

Anytime your Model 3 needs to be towed, it must be done on a flatbed tow truck or using a wheel lift and dolly. All four wheels MUST be off the road. You can't just put it in neutral and drag two wheels. As soon as Transport Mode detects the wheels are moving at a rate faster than 5 MPH, it will sound an alarm and put apply the parking brakes. Ouch!

Please read the section of the owner's manual on **Transporting the Model 3** and make sure your tow truck driver reads it, when the time comes. Pay special attention to the part about the location of the Tow Hook, how to use it, and how to put your Model 3 in **Transport Mode** to disengage the wheels so it can be winched onto the flatbed. We have printed these instructions out, laminated them and keep them in the car so we can just hand them to a tow truck driver should the need arise.

CAR WASH

There's a whole section of the owner's manual devoted to cleaning your Model 3. Here are the highlights:

- Hand washing is best.
- Before you start, remember to turn off the automatic windshield wipers.
- Use cold or lukewarm (not hot) water and a good quality car shampoo.
- If you use a pressure washer, don't direct it at the door seals, window seals or any of the camera housings.

If you're going to take your car to a car wash, make sure it is a TOUCHLESS WASH. This is about more than the paint job, it's about the camera housings and other parts that can be damaged by spinning brushes, etc. Most importantly, if you are using a car wash that drags the car through using wheel chocks, there are several things you should know.

First, it's possible this could damage your rims. As I mentioned in the section about TIRES, the Model 3 rims extend beyond the sidewall of the tires. This means they will likely come into contact with the guide rails in this type of car wash. If you're still intent on doing this, you must put the car in NEUTRAL and **YOU MUST REMAIN in the car while it goes through the wash**.

If the car senses there is no one in the driver's seat, it will immediately shift into PARK which will lock things up inside the car wash and you really don't want to do that. (IMPORTANT NOTE: removing your seatbelt signals to the car that you are preparing to leave the vehicle and it immediately puts the car into PARK. Do NOT remove your seatbelt while going through a car wash. Your car will go into park, get stuck and they will have to shut off the car wash in order for you to drive out. Remember, you drive a Tesla. Your actions reflect on us all.)

TESLA RANGERS

If you haven't figured it out by now, Tesla is in auto industry disruption mode. This company is changing the way cars are made, fueled, bought, driven and now even how they are maintained and repaired. Once you *own* a Tesla it will become clear how these changes benefit all the parties involved. But initially it might seem like there are gaps in some important support mechanisms.

I rarely go two days without encountering a forum post complaining about the terrible Tesla Customer Service. Owners report calling service centers and never get connected to a human or if they do get a human it is not someone who knows the answer or has the power to help them. Yep, I agree. This would be very frustrating, especially when dealing with a car problem.

The good news is that Teslas normally don't have a lot of problems. The other good news is that Tesla seems to be addressing the problem with an innovative concept of Tesla Rangers. This is a mobile fleet of Tesla mechanics in tricked out trucks, vans and converted Model S sedans who come to you and can repair 75% of problems with your Tesla.

The awesome Tesla Ranger service https://electrek.co/guides/tesla-mobile-service/

WHY IT'S GREAT...

The Rangers boast that if the repair needs to happen at your place of business, they can locate your car in the parking lot, move it if necessary and make the repair without you even having to leave your workstation. Now tell me that doesn't sound GRAND.

Being able to dial in and reach an actual human by phone appears limited at Tesla. *(It's possible the 80's called and demanded phone calls back.)* So, the best new way to schedule service is from your MOBILE APP. You will be contacted via email or text (whatever method you have already approved). At that point you can request Mobile service if it's available in your area. The service person will advise you and switch you from the service center to the next available mobile appointment. Or, at least this is how it worked when we scheduled a tire rotation.

This probably goes without saying, but if your car is undrivable or experiencing a dangerous problem, don't wait for a mobile service appointment, call Roadside Assistance immediately and let them take it from there. (NOTE: Posting about your problem on facebook will get you a lot of sympathy, commiseration and maybe even some suggestions. But if it really is an emergency Roadside Assistance should be your first call.)

7

SAFETY

"When something is important enough, you do it even if the odds are not in your favor."
—Elon Musk

Typically, awards are the bane of an industry because determining excellence is so subjective...unless you're talking about automobiles and safety. Once you go there, stringent testing, government regulations and consumer aware campaigns demand data and results and the awards proffered actually mean something.

The Tesla Model 3 basically took home ALL the Oscars when it secured a 5-STAR safety rating from NHTSA (National Highway Traffic Safety Administration) in OVERALL RATING as well as FRONTAL, SIDE and ROLLOVER crash tests. And it scored the "lowest probability of injury" of any car ever evaluated by the U.S. New Car Assessment Program."[22]

More recently, EURO NCAP (European New Car Assessment Program) concurred and also ponied up 5-STARS for the Model 3 in addition to granting the highest award made to date for the Model 3s SAFETY ASSIST features.

While some of these features aren't exactly new to vehicles of this caliber, what is new is Tesla's development strategy. They combined data collected from the Sensor Suite of every Tesla vehicle made since 2016 with the data of billions of inputs from actual drivers navigating all sorts of conditions to create this suite of protections.

The Tech of Tech YouTube (18:00)
How the Tesla Model 3 became the World's Safest Car

If you weren't sure about owning a Tesla Model 3 *before* watching this video, I believe you will be after. Pay special attention to explanation of how the glass roof is an important aspect of the Model 3 safety. With social media you're on your own, however.

LANE DEPARTURE AVOIDANCE

Should you venture close to or actually out of your lane—risky on both social media AND driving—the Model 3 is capable initiating a steering intervention, this means it will literally take control of the wheel and move you back into your lane.

HAVE IT YOUR WAY:

Go to the DRIVING CONTROL menu to select your preference. This setting will remain until you manually change it.

OFF	No lane departure warnings.
WARNING	If the Model 3 detects a front wheel encroaching on a lane marker without a corresponding turn signal, the steering wheel will vibrate and a warning will appear on the screen.[23]
ASSIST	With this option selected you get steering wheel vibration, a visual warning...and *Wait...you also get*...the Model 3 taking control to steer you to a safer position within your lane.

COLLISION WARNING

There are many warnings about all the various road and weather conditions that could defeat this important safety feature so this is one of the sections you should actually read and pay close attention to in the Tesla Owner's Manual. But for the sake of discussion, COLLISION WARNING was designed to assist the driver in the following manner: provide visual and audible warnings when one or more of the sensors on the vehicle detect the imminent risk of frontal collision. The car will apply AUTOMATIC EMERGENCY BRAKING to reduce the impact of a frontal collision and reduce acceleration if an object is detected in the immediate driving path. In other words, the Model 3 tries very hard not to hit anything or anyone.

SPEED ASSIST (part of Autopilot)

The Model 3 detects the speed limit of the road or highway you are on via GPS data. What it does with this information is up to you and how you configure SPEED ASSIST on the DRIVING PREFERENCES menu > AUTOPILOT > SPEED LIMIT.

OFF	Model 3 keeps the information to herself and leaves you to your own devices.
DISPLAY	Displays a speed limit sign on the DRIVING PANE (under Range).

PRO TIP:

Tapping this sign will adjust the TACC speed of the vehicle to the posted speed limit.

CHIME Sounds an alert whenever you exceed the posted speed limit.

HAVE IT YOUR WAY:

Maybe you don't want to be the only A-hole across five lanes of traffic driving at exactly, spot-on the speed limit. The Model 3 has your back and your reputation by offering a setting where you can choose speed limit OFFSET. Go to the DRIVING PREFERENCES menu> SPEED LIMIT > choices are

ABSOLUTE or RELATIVE. Absolute means you select an actual speed and relative means you offset your speed by a number (positive or negative) that you choose.

8

THE GEEK STUFF

"The path to the CEO's office should not be through the CFO's office, and it should not be through the marketing department. It needs to be through engineering and design."

—Elon Musk

HE SAID:

I supported mainframe computing systems for almost 40 years. Believe me when I tell you that ALL computers need rebooting occasionally. If you see some odd behavior on your touchscreen or experience other problems—especially after applying a software update—you might want to do a **touchscreen reboot**. I've done this while driving (and others on Tesla forums have mentioned doing the same). But you should probably do it while parked just to keep your sphincter-factor in check.

REBOOT

DEPRESS both SCROLL BUTTONS on the steering wheel at the same time and HOLD them in until the touchscreen goes dark. RELEASE and wait (about 40 seconds) for the Tesla logo to appear on the touchscreen. Eventually the DRIVING PANE and NAVIGATION PANE, with all the usual controls and indicators, will appear.

On rare occasions, drivers have had their touchscreen black out while they are driving. If this happens, don't panic. You can still drive the car. The steering, accelerator and brakes work independently of the screen. If the screen doesn't come back in about a minute or so, you can just pull over and do a reboot.

There's a mantra pilots try to remember in an emergency: *don't forget to fly the plane*. There are a lot of features and automation on the Model 3. It's easy to get confused and maybe even a bit panicked when something doesn't behave as expected. Just remember, it's still just a car. No matter what else is going on, you can always use the steering wheel, accelerator and brake to make it do what you want. Also, tapping the brake or turning the steering wheel will take back control from AP or TACC features. When all else fails drive your Model 3 like any other car.

POWER OFF

DRIVING PREFERENCES menu > SAFETY & SECURITY > POWER OFF

This turns the car OFF and is considered a **hard reboot**. The Model 3 powers back on again by pressing the brake pedal. NOTE: Power Off takes a full two minutes to complete so you must wait at least that long before powering up again.

SOFTWARE HANGS, LAGS & (maybe) QUICK FIXES

You can't fix your Model 3 if there's really something wrong with it but it's possible the problem you're seeing has an easy solution. Here are some simple things you can check before calling support or embarrassing yourself on the forums.

- If you're having trouble with the Mobile App or if your phone fails to unlock your Model 3, first try nudging the Bluetooth on your phone by either turning it off and back on or activating airplane mode, then back on.
- If a Bluetooth reset doesn't work, next try rebooting the phone. This is actually a better solution than force quitting the Tesla app and restarting it.
- If you're still having problems, check your phone storage. Users have reported connection issues when their phone storage was

nearly full. After clearing out some space and rebooting the problems went away.

- If your problem has to do with the touchscreen display—or any data presented on it—perform a **touchscreen reboot** (see above).
- If a reboot doesn't resolve the issue you can try a complete power down and restart of the car. DRIVING PREFERENCES > Safety & Security > POWER OFF. The touchscreen will immediately turn black but you may have to wait for 30 seconds or so for the power off to complete. Restart the car by pressing the brake pedal.
- Last resort: uninstall Mobile App and reinstall, then reconnect Mobile App to the car.

If none the above steps resolve your issue, it may be time to contact Tesla service or, if it isn't something directly affecting your ability to drive the car you can schedule service from the Mobile App.

BUG REPORT

When an ICE car acts up you have to take it to the dealership. Once you get there, you try to describe what went wrong and you just know that it won't do it again while the service rep or mechanic is driving it. It's so frustrating. Especially, if the problem is frequent but intermittent. You get the feeling that you will have to wait until it fails completely before you can get it fixed.

Model 3 has a better option. Say you're driving and the car does something you think is a software bug. Maybe you're using Auto Lane Change and midway through it jerks the car back into your lane for no apparent reason. You can send a real-time BUG REPORT right then and there, while you are driving. TAP the MICROPHONE ICON (on the touchscreen) or PRESS the RIGHT SCROLL WHEEL say: "Bug Report" immediately followed by a very brief description of the problem. The car will gather information from the data recorder, GPS, a snapshot of the touchscreen, etc. and send it all to Tesla along with your description.

I'm not saying they will respond and address the problem with your car specifically, but they do review these bug reports and it's one of the ways the development team finds and fixes issues in the software. Some drivers report that they actually did get an email from the Customer Care Team acknowledging the bug report.

EVENT DATA RECORDER AND TELEMATICS

Model 3 is equipped with an **Event Data Recorder (EDR)**, similar to the black box in a commercial airliner. In the event of a crash or near crash (or even hitting an obstacle in the road) the EDR will record about 30 seconds worth of data regarding the status of various systems in the car, speed, acceleration, whether seat belts were fastened, etc. The data is collected to help Tesla understand how the various systems performed and where improvements might be made. Although this data doesn't contain any information identifying the driver, it's possible it could be combined with information in an accident report and used in criminal or civil proceedings.

Your car also contains a **telematics system**. This is not specifically related to crash-type events. It gathers information from different systems throughout the car and stores it for use by Tesla technicians during service and troubleshooting, etc. The telematics system also transmits this data to Tesla periodically, all in the name helping to ensure your car is properly maintained. The telematics system is also used to manage software updates, charging reminders and (probably) Supercharging.

TESLA CONNECTIVITY

Your Model 3 connects to the outside world in two ways: using Wi-Fi when it is near a known network and through the cellular network when it's not. The status of this connectivity is reflected in the standard Wi-Fi ICON on the status bar at the top right of the touchscreen.

The cellular connection is pre-configured when you purchase your car. In the USA this connection is through the AT&T cellular network but you won't see any bills from them. They've made a deal directly with Tesla. Tesla will only charge you if you elect the Premium Connectivity package after the first year of ownership.

Standard connectivity includes basic maps, navigation, music and media playing over Bluetooth with software updates over Wi-Fi. Some safety updates will come in over the car's cellular connection, as well. But if you purchased the Premium Interior, you will also get the Premium Connectivity package for FREE for the first year of ownership. This adds satellite map views and live traffic visualization. It will also allow you to stream music and other media from the Tesla network and even adds a web browser to your touchscreen tools. (This is useful for many things, including running **A Better Route Planner** while you are on long trips.) After the first year, Premium Connectivity comes at a price. At the time of this writing, it looks like the price in the U.S. is about $100 per year. By then, you will have had a year to determine whether it's worth it.

Do I think it's worth it? Honestly, the only thing in the Premium package that I would truly miss is the web browser. I don't find satellite views to be more helpful in navigating than standard map views and, while the traffic visualization is somewhat helpful, I'm not confident in its accuracy. I've driven by interchanges that were practically empty of cars but the map showed them as being congested. This may not be Tesla's fault or even Google's. I'm not sure what the source of their traffic data is or what kind of lag times are involved. I only know I have seen these inconsistencies and it affects my evaluation of this particular feature.

As for the streaming media, well I really prefer to stream from my phone through Bluetooth. I already pay for a streaming service there and I have compiled playlists and podcast subscriptions, etc. So, Tesla's streaming services may be fine but I just don't need them. The web browser, however, comes in handy in many situations. Particularly when I'm sitting at a Supercharger getting juiced up and want to surf the web on a big screen instead of my smart phone. Version 10 of the software added (among other things) the ability to stream Netflix and YouTube when vehicles are stopped.[24]

A Wi-Fi connection is necessary for most software updates so you'll want to connect your home Wi-Fi network to your car as soon as you get it home. This works amazingly similar to the way it does on your smart phone. You simply touch the Wi-Fi icon on the status bar at the top right of the touchscreen and it will search for and display the Wi-Fi networks within its range. Select yours and enter the password when prompted then touch CONNECT.

Just like your phone, your Model 3 will remember this network and connect to it any time it's in range. You can configure other Wi-Fi networks when you are near them, as well. It will remember them all and connect to them when in range. If two are in range at the same time, it will connect to the one with the strongest signal. Again – just like your phone.

NEURAL NETWORK

One of the most amazing things about the Model 3 is Autopilot and the promised Full Self-Driving capabilities. Other car manufacturers are working on this but there is little doubt that Tesla has a huge advantage because of the way the computer systems in their cars are built and trained.

Without getting into technical details that I am not qualified to talk about, it is sufficient to say that the computers run something called a neural network (computer simulation of a human brain) that is trained using data from the real-life driving experiences of nearly every Tesla driver in the world.

Your Model 3 and nearly every other Tesla vehicle on the road—several hundred thousand of them—are constantly uploading data about the real-life driving situations Tesla drivers encounter on the road. Anytime you make a real-time BUG REPORT, described in Chapter 8, or simply take over control from Autopilot in a driving situation, data about that event is transmitted over the network to Tesla. If they determine that the car reacted in a way it should not have, they will likely send out a request to the entire fleet of all Tesla cars and solicit data about similar situations.

Tesla will feed all of this data to the neural network so it will learn to better evaluate that situation and respond more like human drivers and AP drivers who handled it successfully. This results in algorithms (computer-generated solutions) which are slightly improved. These solutions are then deployed back out to the fleet but are NOT yet used to drive the cars. Instead, this new code is run on a **shadow processor** that runs in parallel to the actual one driving the car. Now, each time this situation occurs for any car in the fleet, the shadow processor makes a new decision. Some of these decisions will be right and some wrong. This data is sent back in to Tesla and is again used to retrain the neural network and produce new, improved algorithms. This cycle likely occurs several times before the results are deemed acceptable for the new algorithm to be deployed for actual use in driving on AP.

As you can see, the more you drive—the more we ALL drive—our Teslas, the smarter they become. Tesla's neural network currently has data from over two billion hours of actual driving in its database. That's why Tesla is so far ahead in the race to true autonomous driving.

As much as we enjoy driving our Model 3, you should know that Elon's endgame is a driverless car. In fact, Elon is racing to a time when humans will no longer be permitted to drive cars because they simply aren't as good at it as computers. If this alarms you, it shouldn't. There are about 1.3 million deaths every year due to car crashes. Imagine if that number were 90% smaller. Wouldn't that be worth giving up the fun of driving? Besides, think of all the things you could do while being driven to and from work every day.

CHECK OUT:

The 911 on Tesla's Fleet and Neural Network:
TOPSPEED: "Tesla Autonomy Day 2019 – Full Self Driving Autopilot – complete investor conference Event"
YouTube (2:34:59)

TOYS, GAMES & PRANKS

Tesla is not your grandfather's car manufacturer. They don't just stick a radio in the dash and say, "Entertain yourself." They have a whole lot more imagination than that. This is apparent when you open the ToyBox and look at all the things you can play with while you're charging up or...well, there's nothing wrong with sitting in the car in your own garage playing video games for hours, right? Right?

If you tap the Tesla "T" at the top of the touchscreen a popup appears which shows a representation of your car, the odometer, VIN# and installed software release identifier. If you wait a few seconds, this display shifts down to reveal the ToyBox, containing a number of Easter eggs from which you can do everything from wiring passenger seats with electronic whoopie cushions (**Emissions Mode**) to changing the landscape on the navigation pane to resemble mars. There's even a picture of an old-style arcade game and if you tap on this, you will be presented with a number of games you can choose from.

Cuphead described as a classic run and gun action game uses an Xbox controller with a USB cable. **Beach Buggy Racing 2** (reminiscent of Mario Kart) uses the steering wheel on the car and brake pedals to drive in the game, **Chess** for the more sedate among us and a number of old-style Atari games for the old guys like me. It's a great way to pass the time while charging.

The shortcut to the ToyBox is an icon on the **app launcher** but if you just want to go straight to the games, choose the icon to the right of it labelled "Arcade".

9

THE CUTE STUFF

"I do think there is a lot of potential if you have a compelling product and people are willing to pay a premium for that. I think that is what Apple has shown. You can buy a much cheaper cell phone or laptop, but Apple's product is so much better than the alternative, and people are willing to pay that premium."

— Elon Musk

You can say... "It's just a car!" And while this is true...it's also more than that. After nearly a year of reading various Tesla forums and watching Tesla YouTube videos, it's clear that most Tesla owners are as giddy as I am about our car/*notjustacar*. People regularly report verbally addressing their cars in the morning upon waking them up. (Technically, you're waking up the software, but you get the idea.) In the grand tradition of "Herbie" and "Kit" from Knight Rider, the Tesla Model 3 is the perfect vehicle to be anthropomorphized. Begin by officially giving your car a name.

TAP the TESLA LOGO on the top of the touchscreen to bring up the NAME DIALOG BOX. From this point on all messages sent to your Mobile App from Tesla will identify your vehicle by his or her name.

Technically, you don't have to name your vehicle, but I think you will want to.

Where Helva Pearl came from. Well, Pearl is an expression of her gorgeous pearlescent paint. While Helva came from one of my favorite sci-fi books, *The Ship Who Sang* by Anne McCaffrey. The story is about Helva, who was born human but with a deformed body. The custom in this society was to save the brain and essence of the human but discard the body. Brains were then schooled, programmed, and implanted into the sleek titanium body of an intergalactic scout ship and sent out to patrol the galaxy along with a human partner. The ship has human qualities and a personality and always chooses the human partner. Gifted with the voice of an angel and being virtually indestructible, Helva XH-834 anticipated a sublime immortality...but then one day she fell in love! If you read up on Elon's plans for the Neuralink[25]...it sounds like one day there could be a real Helva. But for now, she lives in my garage.

CARAOKE

Sing-along with Tesla! Road trips were totally made for sing-a-longs. Am I right? Note: Lyrics display on the screen when PARKED OR for use by a passenger. Access this feature from the MEDIA MENU. TAP microphone to toggle vocal track on or off.

ROMANCE MODE

Late night parking at the local make out/hang out doesn't have to just be for teenagers. Try it in your Model 3 and heat up the ambiance with ROMANCE MODE. Touch the Tesla Logo and select the fireball. It is complete with sound effects and heat.

PRO TIP:

To make Romance Mode extra cozy add seat heaters but skip the s'mores—too many crumbs. Or, if you're in a silly mood combine with EMISSIONS MODE. I'll let you try that and find out.

TESLA THEATER

Tesla Theater lets you watch Netflix, Hulu and YouTube movies and videos when your car is in PARK. Access this wonderful new feature via the APP LAUNCHER > ENTERTAINMENT.

DRAWING TABLET

There's even an Etch-a-Sketch style drawing tablet on the screen to delight artists young and old. TAP the Starman from the ToyBox to access.

DRIVING GLOVES

Hear me out...the glass roof has both infrared AND UV protections built in. But I am fair-skinned and the backs of my hands are prone to unsightly dark spots from the sun. My dermatologist told me it was from driving and not putting sunscreen on my hands. The full expanse of windshield in the Model 3 makes for nice visibility but it also exposes my hands to more regular, unfiltered sunlight than is probably good for me. So, I invested in a cool pair of pink driving gloves that I keep in the console.

MATCHING HANDBAG

Pearl's luxurious white interior is a thing of beauty and none of my black, brown, red, navy, two and/or multi-colored handbags did it justice. So, of course this required a new handbag purchase. You can see a photo on our website: http://TeslaModel3Guide.com

TESLA EASTER EGGS

Elon must have a really good sense of humor because some of these are really fun. In addition to the game and gag features in the Tesla Toybox (TAP the Tesla logo on your touchscreen) there are a couple of Easter Egg events that have been discussed in various forums. We'll tell you how to access them, then it's up to you to find out what they do.

RAINBOW ROAD: While driving in Autopilot DEPRESS the RIGHT STALK FOUR TIMES in rapid succession. Then be amazed. Repeat action to turn it off.

BACK TO THE FUTURE: If you remember the movie with Michael J. Fox and Christopher Lloyd in order to turn Doc Brown's DeLorean into a time machine, he needed 1.21 gigawatts of power. Drive around in your car until the range indicator (set to display miles) ticks over to 121 miles. The Easter Egg occurs on the screen of your Mobile App (not in the car.) TAP on the battery at the top of the Tesla Mobile App to launch the Easter Egg. Then explore and be amazed.

These are the ones we know about now. There are probably more. Let us know if you find any we missed.

SHOULD YOU BUY A CAR OFF THE INTERNET

Tesla isn't the only one trying to streamline or *disrupt* the traditional car buying process. If you buy a Tesla online and don't like it, you can return your car in 7 days or 1,000 miles, whichever comes first. But should you? That is the question?

WHAT WE DID

We live close enough to a Tesla Dealer and were able to test drive before ordering. But also, we actually met on the internet...in a writer's group. And we bought our house off the internet (without seeing it in person first). So, maybe we aren't the best people to advise in this area. My initial advice is do whatever you have to do and just get the dang car. The only potential problem is if there is something wrong with the car, like a defect or flaw. I have read of problems where people have returned the car, hoping to replace it with one without a defect. But then had trouble getting the money/loan turned back to zero.

Or, I've heard of Tesla delivering cars right to people's doors, with a delivery team in place to set it up. This is great...unless the car needs to go back. What then? Do they pick it up or do you have to drive it back? I think most likely direct delivery will be fine and you will love the car. Just be very **clear** about **all** your options and who to contact in the event of a problem.

COST VS. RESALE: IS IT WORTH IT

The Model 3 is clearly a luxury vehicle with a luxury price tag but ARK Invest did a study comparing the cost of ownership over a three-year period with that of buying and owning a Toyota Camry. Based on Kelley Blue Book Best Resale Value Awards for 2019, the Model 3 was predicted to retain 69.3% of its original sales price, while the Toyota Camry will only retain 48.6% of its cost. Once you factor in the cost of operation over three years the Tesla Model 3 comes out ahead.[26] Tesla FTW!

10

EVERYTHING ELSE

"Persistence is very important.
You should not give up unless you
are forced to give up."
—Elon Musk

TESLA ETIQUETTE: DON'T BE THAT TESLA D*CK! (DBTTD)

Not only are most Tesla Owners really in love with their vehicles but we feel like we're part of something special...something bigger than *hey, cool car*. The Supercharger setup, as well as the myriad of online groups and forums, give Tesla owners an opportunity to band together both online and in person. Tesla owners share information, their favorite tips and tricks along with the latest Tesla mythology and lore. It doesn't take long for the topic of inconsiderate Tesla owners to come up.

Here are some complaints I repeatedly see on various forums.

- Leaving a pile of trash behind at a Supercharging station. Seriously, I've seen the photos—drink cups, food bags, even dirty diapers. I get it, you have a nice car and you want to keep it that way. But WTH, you've stopped spewing emissions and destroying the ozone BUT now you're a LITTER BUG? Get a bag designated for trash and use it. DBTTD!
- A small wave or nod when passing another Tesla is not mandatory but most Tesla owners kind of expect it. If there was a cool kids table for cars, the Teslas would dominate AND they would be nice to everybody. DBTTD!
- When you pull up at a Supercharger, if possible, don't share a charging circuit with another vehicle. Obviously, if there are only

shared spaces available, you're going to do what you have to do. In that case everyone gets it and it's not a dick move. Just be aware.

- On a road trip you'll likely avail yourself of FREE destination charging at various hotels along your route. Many of the larger chains are fully equipped with several chargers for their guests. And, when on a trip, it's sometimes vital to get as much charge as possible to get you to the next leg of your journey.

SHE SAID:

- On our last road trip we checked in, plugged in and crashed. The next morning, we figured out that another Tesla owner came in after us and also wanted to plug in. But they couldn't because the charger was a Tesla Destination charger and the cable was locked in our charging port. The next morning, my husband moved the car right away while I packed us up and it wasn't long before the other Tesla owner came down to plug his car in. He thanked us for moving and estimated he still had time to get his charge in before he had to continue on his trip. We realized that had we known he needed to plug in we could have released the charge cable from the Mobile App without leaving our hotel room. From now on we're putting a sticker with our phone number inside the charge port. This way if someone arrives after us and needs to plug in, they can notify us. We can unlatch, using the Mobile App. And if the hotel offers two spaces per charger, it's possible we wouldn't even have to move the car. Moral of this story: DBTTD—when using a destination charger overnight, leave a note with your phone number somewhere visible.

PRO TIP:

If you're the one who arrives late and someone already has the destination charger don't stress or sweat it, just plan a breakfast stop at a Supercharger to top off both tanks.

- Arrogant attitudes abound on the internet and even though Tesla automobiles are a cool product there are those for whom the car will never be right. Paint, body gaps, leather stitching imperfections and software glitches...oh, forever the software glitches. Basically, there's no end to the negative spew for some people. Now, I'm not saying that anyone should ignore or accept anything less than perfection. But don't go all over the internet dragging Tesla. Yeah, there are some things that aren't perfect. Welcome to the world of mass production, assembly-line craftsmanship and hell, even growing pains. The point is these errors occur on ALL cars ICE and EVs—it's why the recall system was developed. I recently read a very favorable review of a new model of an ICE vehicle. While the review said this was a great car, it also mentioned there were no less than 20 recalls in the first five months of the vehicle release. The point is get your car fixed, get it made perfect, just don't trash-talk Tesla on the internet. DBTTD!

- RTFM—stands for Read The F*cking Manual. Not you, gentle reader. You've made it this far, you KNOW things. So, when a very excited newbie storms the Tesla Forum where you like to hang out and asks a basic question you've seen a hundred times and answered about 50. If you don't want to answer 51 times, it's cool. Just glide past the post and go on your way. In the time it would take you to stop and fire off a snarky RTFM post you could find and share a cat meme instead. Which means you could contribute to saving the planet and lighten up the internet at the same time. So, don't post that RTFM response to an excited newbie's question. It makes you look bad and it's bad for the brand. DBTTD!

TESLA MYTHOLOGY AND LORE

Are the below statements true or am I just making stuff up?

ELON NAMED HIS VEHICLES TO SPELL "SEX".

A: True!

Well, to be precise he had already named the Model S and X and during an interview, someone suggested he should name his third model the 'E' so the line-up of vehicles would spell SEX. Elon's sense of humor is evident in many aspects of his cars and it's something I really appreciate.

Apparently, the story goes that he tried to get a trademark for Model E but it was granted to Ford. Elon made the comment to a group of Shareholders that: "Ford killed SEX." He ended up trademarking Model 3 and expressing it with the three bars to look like an E. Which means he got SEX after all, along with a whole new look to the Brand. With the coming Model Y, he'll have S.E.X. and Y. Y? Because it's cool and we like it! Elon is cool and as Tesla owners we can capitalize on a little bit of that cache. Nobody can stop us.[27]

ELON MUSK IS THE LONGEST TENURED AUTOMOTIVE CEO RIGHT NOW.

A: True!

When Deiter Zetsche stepped down from Daimler in May 2019, Elon went to the head of the class.[28]

GM AND FIAT CHRYSLER PAID TESLA HUNDREDS OF MILLIONS OF DOLLARS.

A: True!

Tesla sold extra zero-emissions regulatory tax credits to GM and Fiat Chrysler which helped those companies meet the regulatory requirements that would have resulted in punitive penalties. Talk about a win/win scenario—they needed a pass and he could use the cash.[29]

TESLA IS MORE THAN A CAR COMPANY.

A: True!

If you've been following along throughout this book you have to admit there is nothing traditional about this car or the company. Tesla appears poised to do for the auto industry what Apple did for smartphones. Also, Tesla Solar Panels, FTW!

THE TESLA MODEL 3 HAS A BIO-WEAPON DEFENSE FEATURE.

A: False!

Both the Model S and Model X have a hospital-grade HEPA filtration system (labeled Bioweapon defense) that can even filter out chemical weapons. The Model 3 air filtration system is not hospital-grade, but still blocks pollen, industrial fall out, road dust and other particles from entering the cabin.[30] Due to size constraints of the Model 3 dash the filtration system is more limited than the ones in the Model S and Model X.

WHY IT'S GREAT...

Many Tesla drivers fleeing the deadly California fires in 2018 praised Tesla's filtration system as helpful during the extremely negative air quality. Tesla also routinely puts out an OTA for cars in evacuation areas that extended their range. And, Tesla has made Supercharging in hurricane areas FREE while evacuation orders were in effect. By comparison, gas stations in the areas were charging outrageous inflated prices. (NOTE: In a bad air quality situation, especially smoke, it is advised to select recirculate air and turn the FAN on level 5.) Tesla's decisions are nothing short of heroic. But how great is it that they had the foresight to design a car where they can be heroic?

MODEL 3 APPS

Every day more apps appear that support Tesla cars in some way. Generally, they fall into one of these categories: apps that report on the car's status, apps that assist in driving (trip planning, finding chargers, etc.) and apps that process data about or from the car.

We include here a few that we think are helpful. I'm sure more will be available by the time you read this. For now, I hope this will give you an idea of the kinds of functions you can get from these programs. Please review each app's features in the app store as we won't be able to fully describe them here and new features may get added by the time this is published.

Oh, one more thing I should mention about all these apps. The only way they can get information about your Tesla is to logon to your Tesla account and get it from Tesla's servers. They do not and cannot access your car directly. That's a good thing. However, in order for the apps to logon to your Tesla account, they will ask you to provide your Tesla userid and password to the app. Most apps will explain that these credentials are only stored on your phone and never on their servers and some apps will even accept a "token" in the place of your userid and password. There are several ways to get a token (including the Tesla Token app listed below) but tokens do expire after some interval (usually 30–90 days) and must be regenerated. So, there are advantages and disadvantages to using the token instead of trusting an app with your Tesla credentials. You'll want to consider this when using apps that offer the choice.

APPS THAT RUN ON A SMARTPHONE:

Stats: For Tesla Model S/X/3 $19.99 (allows Family Sharing)

Possibly the most popular mobile app with Tesla owners. It provides lots of interesting charts and graphs about your driving and charging activity, allows you to compare your driving efficiency to others in the fleet, track how much money you're saving compared to using gas, etc. It also offers many of the same controls available in the Tesla mobile app (lock/unlock, flashlights, honk horn, start/stop charging, etc.) and provides links to the most popular Tesla and Model 3 forums online. It includes smart features regarding climate control, sentry mode and battery prep, as well. Twenty bucks might seem steep but it's a one-time charge compared to subscription rates required by some of the other apps.

Remote for Tesla $9.99 (allows Family Sharing)

This is a nice, compact app that gives you all the basic status information and controls on a single screen. It's perfect if you aren't interested in historical data regarding range, trips or performance against the fleet, but want info on your SoC and the ability to issue standard control commands.

Plus—for Tesla Model S/X/3 $2.49/month

Plus provides the standard status info and controls but also includes **schedules** (need to warm your car every morning in the winter before your commute?) and **camping mode**, which solves the climate issue for folks who want to sleep in a warm or cool (i.e. climate controlled) Tesla without having to leave the headlights on all night.

EV Trip Optimizer for Tesla $0.99/month

EVTO, as it's called, attempts to do a better job of planning your trips than the online Tesla Trip planner or the trip planner that's part of the GPS navigator built into the touchscreen display in the Model 3. It allows for up to 20 waypoints and when calculating drive times and charging stops it takes into account things like weather and wind along with the usual issues of elevation changes and temperature. It also handles multi-day trips more easily than the Model 3 built-in trip planner. You don't have to pay for the subscription but without it you will be limited to trips of 500 miles or less and miss out on some of the best features, such as, allowing you to send the trip route and stops to your Model 3 navigator via calendar entries.

EV Watch for Tesla (free)

Just a basic status/control app but, as the name implies it allows you to display status information about your Model 3 right on your Apple watch and even issue the usual control commands from there.

Toolbox—Remote for Tesla $6.99

Similar to other status/control apps but also comes with an Apple Watch **complication** and allows you to use summon from both the phone and the watch. A word of warning about summon from the phone, however. When you tap the summon icon (the small navigation arrow), a little window appears with options to summon "Forward", "Reverse" or "Abort Summon". Once you choose forward or reverse, a popup appears prompting you to enter your Tesla password before it will perform the action. That's good BUT after you enter it, the popup goes away as does the window that allows you to stop the summon. You must remember which icon you used and tap it again, then choose ABORT to end the summon. Nerve-wracking? Yes. Dangerous? I think so but perhaps they will improve this at some point. Caveat emptor. (**HE SAID**: *for the longest time, I thought that's how to tell the caterer to refill the caviar dish. I thought it was Latin for "caviar empty."*)

APPS THAT ARE WEB-BASED:

TeslaFi 5.00/mo or 50.00/yr

A very popular web application that tracks driving and charging time among other things. It even has an Alexa skill so you can request some info and even execute some commands using your Amazon TAP device. Pay special attention to the Sleep Modes if you use this app. If not set correctly, the app can prevent your Model 3 from sleeping which can cause additional vampire drain. It features a trial period so you should definitely try it out to see what kinds of information it presents that may be of interest to you.

TeslaCam Video WebApp Player (free)

(sentrycam.appspot.com)

The best way we've found to view the camera clips saved to your TeslaCam flash drive folder. It combines all data from a single event and displays all angles simultaneously on a split screen. It even features a **Jump to Event** button to take you directly to the event that caused Sentry to save the video (if it's a Sentry event video) so you don't have to scroll through ten minutes of video to find it.

Tesla Token $ .99

If any of the applications you intend to use allow you to supply a token in place of your Tesla userid and password, this application will acquire that token and allow you to copy it to the clipboard so you can paste it into the app.

TESLA MODEL 3 (must see) VIDEOS:

Search for links to these videos on Google or on our website at www.teslamodel3guide.com!

TESLA	Tesla Unveils Model 3	YouTube (22:44)
ANDY SLYE	Tesla Model 3 Guide: What to Know Before Buying	YouTube (24:52)
	Tesla Model 3 Review: The Truth After 26.000 Miles	YouTube (12:33)
Bjorn Nyland	Tesla Model 3 Interior Seat and Cargo Space	YouTube (32.20)
Now You Know	Tons of Storage - 12 days of Model 3!	YouTube (3:55)
Teslucky	A mother/daughter Tesla YouTube channel. They drive a Model X but interview Tesla Owners at Supercharging stops.	YouTube (so CUTE!)
Wade Anderson	Tesla Social Multiple videos of 41-day road trip to remote locations	YouTube (So Cute!)

TESLA GROUPS AND FORUMS

Tesla Divas	https://www.facebook.com/groups/TeslaDivas/
Tesla Forums	https://forums.tesla.com
Tesla Owners	https://teslaownersonline.com
Tesla Motors Club	https://teslamotorsclub.com/directory/
Tesla Model 3/Y owners	https://www.facebook.com/groups/TeslaOwnersOnline/

TESLA LINKS FOR DATA, NEWS & ACCESSORIES

Cleantechnica.com

Evannex.com

Taptes.com

Teslarati.com

ABetterRoutePlanner.com

TESLA ON TWITTER: (follow these)

@elonmusk

@Tesla

@TeslaDaily

@Teslarati

@TeslaOwnersOnline

@MYMODL3

@TeslaModel3News

@Tesletter

@Cleantechnica

@Scarbo_Author (Sheryl Scarborough)

@EqualEyes1 (Jerry Piatt)

THE TESLA FACTORY TOUR

We haven't been on a tour of the factory, but heard it was amazing. For information and to arrange a tour email: factorytours@teslamotors.com with your information and requested dates. They will get back to you fairly promptly.

CHECK OUT:

WIRED: Elon Musk's Master Plan Culminates in the Tesla Model 3 YouTube (7.00)

ACKNOWLEDGEMENTS:

SHE SAID: This book was written at a time when it seemed like our world was literally melting and burning to the ground. My marker for Summer 2019 will remain Greta Thunberg labeling this the beginning of a mass extinction event! If ever I felt a need to add my voice to an important issue it is now! Hopefully my words will encourage someone to buy a Tesla without fear that the technology is too difficult for them to learn. I immersed myself in the wonder of Tesla over these past months. And, as a result, I have hope again. The Tesla mission combined with the Tesla Owners' community will do that for you. I've always said it takes a village to get an author to the point where there is an actual book. And this book is no exception. Thank you to Yfat Reiss Gendell who taught me how to do this. Huge thanks to Kate Kuch, Laura Moe, Beth Bacon and Marge Engessor for sharing hard-earned self-publishing savvy. Thanks to Brett Lyerla for cover design and to John Garvin at Atomic Cherry for the badass website. Cheers and Happy Trails to all the great Tesla owners (& Tesla DIVAs) I've met and to the forums where we can talk Tesla to our heart's content. Elon, you're the best! Years ago, I swore I would never buy another American-made car. You made me a liar. The Model 3 is a damn sweet ride. Finally, to my partner in all things (even Helva-Pearl) and now also in a book, Jerry Piatt you are my world. Like always, I couldn't have done it without you...and of course, Gizmo, the writer cat.

HE SAID: I am not a writer. That's the first thing I want to acknowledge. Any mistakes you find in this book are probably mine. Having said that, I want to thank my wife for allowing me to put my name next to hers on this book. If you don't know any writers, you probably don't realize how fantastically difficult it is to get one to share credit. But seriously ... it's REALLY difficult. So, thank you to the woman who actually bought and owns the Tesla I get to drive. She's a hardcore Elon fan and a ride-or-die Tesla evangelist from the very early days. I also want to say thanks to Stuart and Lois Fox, fellow Model 3 owners who reviewed the manuscript and gave us extremely helpful notes, finding mistakes and suggesting some things we would otherwise have completely forgotten to mention. Lastly, I want to give a shout out to Alex, an old boss of mine who once laughed at me when I wrote on my performance review that I had excellent communications skills. Suck it, Alex! I wrote a book! What have you done lately?

USE OUR REFERRAL CODE:
Sheryl Scarborough or **https://ts.la/sheryl51287**
Free Supercharging miles for YOU and for US!

Bibliography

*All Elon Musk quotes Kim, Larry *Inc.com* "50 Innovation and Success Quotes from SpaceX Founder Elon Musk" Accessed: September 13, 2019 https://www.inc.com/larry-kim/50-innovation-amp;-success-quotes-from-spacex-founder-elon-musk.html

[1]"All-Wheel Drive, Dual Motor" Tesla.com Accessed: September 11, 2019 https://www.tesla.com/model3

[2]"definition of tipping point" Merriam-Webster Accessed: September 11, 2019 https://www.merriam-webster.com/dictionary/tipping%20point

[3]"Valet Mode operation" Tesla Owner's Manual 2019.16.1.1 Accessed: September 11, 2019 https://www.tesla.com/sites/default/files/model_3_owners_manual_north_america_en.pdf

[4]Shaffer, Leslie "Electric vehicles will soon be cheaper than regular cars because maintenance costs are lower, says Tony Seba" CNBC.com June 14, 2016 Accessed: September 11, 2019 https://www.cnbc.com/2016/06/14/electric-vehicles-will-soon-be-cheaper-than-regular-cars-because-maintenance-costs-are-lower-says-tony-seba.html

[5] Huddleston Jr, Tom "Tesla's Model 3 ranked most satisfying car, more than Porsche or Corvette" CNBC.COM February 11, 2019 Accessed September 11, 2019 https://www.cnbc.com/2019/02/11/consumer-reports-teslas-model-3-most-satisfying-car.html

[6] Harris, David "How Far Do Americans Drive to Work on Average?" Itstillruns.com Accessed: September 11, 2019 https://itstillruns.com/far-americans-drive-work-average-7446397.html

[7] "The Asphalt's Getting Crowded" Facethefactsusa.org January 25, 2013 Accessed: September 11, 2019
http://www.facethefactsusa.org/facts/the-asphalts-getting-crowded-video-/

[8] "Tesla Range Table" Teslike.com Accessed: September 11, 2019
https://teslike.com/range/

[9] @elonmusk "In reply to @likeTeslaKim" twitter.com November 30, 2018 Via Cleantechnica.com Accessed September 11, 2019
https://cleantechnica.com/2018/12/07/how-to-charge-your-tesla-overnight-to-keep-your-battery-healthy/

[10]"Less Than the Cost of Gas" Tesla.com "Less Than the Cost of Gas"
https://www.tesla.com/supercharger

[11] Hanley, Steve "Tesla Batteries Have 90% Capacity After 160,000 Miles, May Last For 500,000 Miles" Cleantechnica.com April 16, 2018 Accessed September 11, 2019
https://cleantechnica.com/2018/04/16/tesla-batteries-have-90-capacity-after-160000-miles-may-last-for-500000-miles/

[12]"Impact Report" Tesla.com "Impact Report"
https://www.tesla.com/ns_videos/tesla-impact-report-2019.pdf

[13]"Controlling Lights", Tesla Owner's Manual 2019.16.1.1 Accessed: September 11, 2019
https://www.tesla.com/sites/default/files/model_3_owners_manual_north_america_en.pdf

[14] "Interior Storage and Electronics", Tesla Owner's Manual 2019.16.1.1 Accessed: September 11, 2019
https://www.tesla.com/sites/default/files/model_3_owners_manual_north_america_en.pdf

[15]"About Autopilot", Tesla Owner's Manual 2019.16.1.1 Accessed: September 11, 2019

https://www.tesla.com/sites/default/files/model_3_owners_manual_north_america_en.pdf

[16]"Traffic-Aware Cruise Control/"Adjust the Following Distance", Tesla Owner's Manual 2019.16.1.1 Accessed: September 11, 2019 https://www.tesla.com/sites/default/files/model_3_owners_manual_north_america_en.pdf

[17]Ferris, Danica J "Tesla Reviewer reveals a secret of the Model 3's stellar sound system" Teslarati.com August 14, 2019 Accessed: September 13, 2019 https://www.teslarati.com/tesla-model-3-sound-system-secret/

[18]"Security Settings", Tesla Owner's Manual 2019.16.1.1 Accessed: September 11, 2019 https://www.tesla.com/sites/default/files/model_3_owners_manual_north_america_en.pdf

[19]"Autopilot Features", Tesla.com "Autopilot Features" https://www.tesla.com/support/autopilot#autopilotfeatures

[20] "Some Newer Cars Are Missing a Spare Tire" Consumerreports.org August 6, 2018 Accessed: September 11, 2019 https://www.consumerreports.org/tires/some-newer-cars-are-missing-a-spare-tire/

[21]Alvarez, Simon "Tesla Model 3 Aero vs Non-Aero Wheels Real-World Efficiency Test" Teslarati.com January 19, 2018 Accessed: September 11, 2019 https://www.teslarati.com/tesla-model-3-aero-vs-non-aero-wheels-real-world-efficiency-test/

[22]"Built For Safety", Tesla.com "Model 3, Built For Safety" https://www.tesla.com/model3

[23]"Lane Assist/Lane Departure Avoidance", Tesla Owner's Manual 2019.16.1.1 Accessed: September 11, 2019 https://www.tesla.com/sites/default/files/model_3_owners_manual_north_america_en.pdf

[24] Palmer, Annie "Elon Musk says Teslas will soon be able to stream NetFlix and YouTube" CNBC.COM July 29, 2019 Accessed: September 11, 2019 https://www.cnbc.com/2019/07/29/elon-musk-says-teslas-will-soon-be-able-to-stream-netflix-and-youtube.html

[25] Rogers, Adam "Here's How Elon Musk Plans to stitch a Computer into Your Brain" Wired.com July 17, 2019 Accessed: September 13, 2019 https://www.wired.com/story/heres-how-elon-musk-plans-to-stitch-a-computer-into-your-brain/

[26]Korus, Sam "Opinion: This math shows Tesla's Model 3 is cheaper to own than Toyota's Camry" Marketwatch.com April 19, 2019 Accessed September 13, 2019 https://www.marketwatch.com/story/this-math-shows-teslas-model-3-is-cheaper-to-own-than-toyotas-camry-2019-04-17?utm_source=zergnet.com&utm_medium=referral&utm_campaign=zergnet_4021341

[27] Truong, Alice Fast Company June 3, 2014 "Elon Musk Wanted Tesla Model Names to Spell The Word "Sex" Accessed: September 13, 2019 https://www.fastcompany.com/3031447/elon-musk-wanted-tesla-model-names-to-spell-the-word-sex

[28]Fortuna, Carolyn CleanTechnica.com "When It Comes To Automotive CEO Longevity, Tesla's Elon Musk Is Now King" July 15, 2019 Accessed: September 13, 2019 https://cleantechnica.com/2019/07/15/for-automotive-ceo-longevity-teslas-elon-musk-is-now-king/

[29]Rapier, Graham Markets Insider.com "Tesla has made hundreds of millions of dollars selling tax credits..." June 3, 2019 Accessed: September 13, 2019 https://markets.businessinsider.com/news/stocks/tesla-sold-carbon-emissions-credits-to-general-motors-fiat-chryser-2019-6-1028250175

[30]"Cabin Air Filter", Tesla Owner's Manual 2019.16.1.1 Accessed: September 13, 2019 https://www.tesla.com/sites/default/files/model_3_owners_manual_north_america_en.pdf

AUTHOR BIOS:

SHERYL SCARBOROUGH'S writing has appeared in just about every medium imaginable, from young adult mysteries, television scripts, series concepts, comic books, children's books, business plans, technical writing, magazine articles and online how-to articles. She has even written theater and restaurant reviews in exchange for great seats and free food. She's happiest sitting at her computer creating something new, which is why she loves her Tesla Model 3 so much...it's almost like sitting at her computer only better because she gets to drive around! Find her as **Scarbo_Author** on Facebook, twitter and Instagram or write to her at SheSaid@TeslaModel3Guide.com

JERRY PIATT worked for more than thirty-five years in mainframe computing. For the last twenty years of his career, he worked as a software developer writing assembler and C++ code for an IBM branded storage performance monitor. He's a fan of the Maker movement, dabbles in digital electronics and loves reading thriller fiction and watching horror movies. These days, he's happily retired and living with his lovely wife in the great Pacific Northwest. Find him on Facebook, twitter, Instagram or write to him at HeSaid@TeslaModel3Guide.com

Index

Made in the USA
San Bernardino, CA
27 December 2019